A World List
of Mammalian Species

A World List of Mammalian Species

Second edition:

G.B. Corbet and J.E. Hill

British Museum (Natural History)

British Museum (Natural History)

London

A World List of Mammalian Species

Second edition 1986
Reprinted 1987

ISBN 0–8160–1548–1   U.S.A.

Library of Congress number   86–80583

British Library Cataloguing in Publication Data
Corbet, G. B.
A world list of mammalian species—
2nd ed
1. Mammals—Classification
I. Title   II. Hill, John Edwards
599′.00216   QL708
ISBN 0–565–00988–5

Printed in Great Britain
at the University Printing House, Oxford

Contents

CONTENTS

CONTENTS

CONTENTS

CONTENTS

Introduction

After birds, mammals are the group of animals whose diversity has been most intensively studied. But the attention of zoologists, whether concentrating on structure, ecology, behaviour or physiology, has been very unevenly distributed amongst the four thousand or so species that make up this assemblage. Some, like the chimpanzee, have had volumes devoted to every aspect of their structure and lives; at the other extreme species have been described on the basis of such flimsy evidence as to leave their very existence in doubt.

What is a mammalian species?

In devising a classification of mammals, or any other group of animals, the basic task is to detect, amongst the bewildering variety of individual animals, those genetically independent groups that we call species and that the animals themselves would recognize as constituting 'us' as distinct from 'them'. This concept of the species is not absolutely rigid and clear-cut but it is based upon the fact that individuals within a geographically coherent group show a degree of resemblance to each other that arises from a common ancestry and allows them to interbreed freely, thereby disseminating their characteristics within the group. The difference between members of such a species and members of the most closely related species occupying the same area may be conspicuous or obscure it is the absence of intergradation rather than the magnitude of the difference that enables species to be defined. Given adequate information there is rarely difficulty in defining the species at any one locality, but the concept of the species becomes less precise when we consider geographically separated populations and have to speculate as to whether any differences between them are sufficiently slight that they would interbreed freely if reunited.

In some groups of mammals, for example the ungulates and carnivores, it is unlikely that many completely new species await discovery and our delimitation of known species is unlikely to change much. However some uncertainties are almost bound to remain since species are continually evolving and it is almost inevitable that some pairs of populations have diverged to just such an extent that it is rather arbitrary whether they are deemed to have crossed the species threshold and been irretrievably set upon separate evolutionary pathways. Amongst the antelopes for example the four principal forms of oryx in Africa and Arabia, all geographically isolated from each other, have at one extreme been considered races of a single species, *Oryx gazella*, but are more often treated as three species, as here (p.139), or even four.

Amongst the small mammals the situation is very different. Several completely new species, mainly bats and rodents, are discovered each year. Known species are sometimes found, on closer inspection, to comprise two or more separate but very similar species – 'sibling species' – whose genetical independence may sometimes be confirmed by the degree of difference in their chromosome complements or by biochemical studies. However, because these techniques cannot usually be applied to existing collections in museums, it frequently happens that the samples available for study are so small or so unrepresentative of the species as a whole that the results are equivocal and it may not be practicable to use them in compiling a list such as this. On the other hand any such proliferation of known species tends to be counter-balanced by the discovery that other 'nominal species', i.e. forms that have been described and named as species on the basis of inadequate samples, are in fact only variants of other, better known, species.

The scope of the list

This book is an attempt to present a comprehensive list of all living species of mammals as far as current knowledge allows. Recently extinct species are also included, marked †, if their external appearance is known from preserved specimens, illustrations or descriptions, e.g. the quagga, an extinct zebra from southern Africa, but not if they are known only from skeletal remains. (An exception is made in the case of the beaked whale *Mesoplodon pacificus*, at present known only from two skulls but quite likely to be surviving.) A number of 'nominal species' are also omitted if it seems unlikely that they represent independent species even if they cannot be confidently allocated to a particular known species (usually because they are based on inadequate descriptions and the original specimens are lost or poorly preserved). The best place to find more information on any species is generally the most recent regional work, for example those listed on pages 227 to 229. For very few groups of mammals are there comprehensive descriptive works that ignore national frontiers.

Higher classification and sequence

It is not intended that this book should be considered an original source for the higher classification of mammals, i.e. classification above the level of genus. The classification used is based, like those in most recent compilations, on that of Simpson published in 1945 (see p.227), adapted to take account of more recent work. The result is a compromise between keeping a well-known classification in spite of the fact that subsequent work has shown some aspects of it to provide a poor indication of relationships, and recent alternatives that have not yet had time to be adequately tested. Ranks other than order, family and genus have been avoided except in the Chiroptera and in a few other cases where they are particularly clear-cut and usefully divide large groups, e.g. in the rodent family Muridae where subfamilies have been used.

We have likewise followed the *sequence* used by Simpson unless the requirements of a more recent classification have directed otherwise. At the level of the orders we have adopted the principal conclusions of the phylogenetic classification of McKenna (see p.227) but have minimized changes of sequence and rank. The main effect of this is to separate the order Edentata from the rest of the placental mammals and within the remainder to recognize the Macroscelidea and the Lagomorpha as comprising a pair of distinctive groups. There are severe limits on the extent to which a linear sequence can reflect the tree of relationships uniting a multitude of groups. We have therefore adopted a simple alphabetical sequence of species within each genus. This has the disadvantage of separating many closely related species but any simple alternative could equally mislead in implying, by juxtaposition, affinities that were not intended. The disadvantages of a non-alphabetical arrangement of genera and higher categories are alleviated by the provision of an index.

Number of species

In quoting the number of species in an order or family the form 'c. 23 species' is generally used, meaning that the number of species *listed* is exactly 23 but that there is sufficient uncertainty about the status or content of some of them to make it likely that this number will be revised as the species become better known, or will remain unstable because of species that are marginally distinct. Absence of the 'c.' implies that the species are clearly defined and the number is unlikely to change.

Nomenclature

No authors nor dates are given with the scientific names – these can be found in the checklist by Honacki *et al.* (see p.227) and in many of the other references quoted. Except when dealing with problems of nomenclature, or with recently described species, it is usually quite unnecessary to quote the author or date of the name of a mammal. Again with the exception of recently described species, the original citation is rarely the best place to go to find a useful description of the species.

The name of the genus is followed, in parenthesis, by any synonyms that are frequently used in recent literature. Zoologists should note that these 'synonyms' are simply alternative generic names that are commonly used for all or any of the species listed in the genus – it is not necessarily implied that they are formal junior synonyms of the generic names used. The same applies to synonyms of species names – these most often represent forms that have recently been considered separate species but are here included in the other, earlier-named, species.

Vernacular names, in English only, are given if they are well established or if they are used in a major work on the region or group concerned. We have invented a few names, especially in cases where an obvious name suggested itself by analogy with a related species that already had an accepted English name. We have also enlarged some well established names to make them unique, e.g. Eurasian red squirrel, without implying that the expanded version need always be used. However, many vernacular names, especially of bats and rodents, are only unambiguous when used in a local context.

Geographical range

The aim has been to be as precise as possible within the severe limitations of space, helped in many cases by an indication of the vegetation zone or zones concerned.

Small islands are generally ignored unless they comprise a particularly significant part of the range. The form 'Uganda, etc.' is used to mean that the range is centred on Uganda but extends slightly into adjacent countries. In south-eastern Asia the expression 'Burma – Java' implies presence in Malaya and Sumatra but not Borneo and Philippines unless these are specifically mentioned. Parentheses ( ) are used to indicate a minor component of the range, especially in the case of genera or higher groups. For example in the case of the opossums of the family Didelphidae 'S, C America, (N America)' reflects the fact that there are many species in south and central America but only one in north America. Brackets [ ] indicate populations introduced by man, but are generally used only in the case of very extensive and well established populations remote from the original range, e.g. coypus introduced from South America into Eurasia and European rabbits in Australia. Introductions are ignored in giving the range of higher groups. In the case of a species in which a major diminution of range over the last hundred years or so has been well documented current range is given but areas formerly occupied are indicated where practicable. No attempt has however been made to indicate earlier ranges known only from fragmentary information or subfossil remains. Extinct species are marked † against the scientific name as well as 'extinct' preceding the range.

Habitat and ecology

This information is not intended to be comprehensive but is used especially to define the range more precisely. Where habitat is constant for a genus or higher group it is not repeated under each species. Expressions such as 'forest' or 'steppe' are intended to define the gross biotic zone that the species occupies rather than the habitat selected by individuals. Generalized descriptive terms for structural vegetation types are used in preference to those with local or floristic connotations, e.g. 'scrub' rather than 'chaparral'. The principal terms used for vegetation types can be defined as follows:

Forest – dominated by trees. Qualified as closed, open, rain, dry, deciduous, evergreen, coniferous.

Woodland – sometimes used here for open forest but generally avoided because of ambiguity. It can also mean secondary forest or small scattered segments of forest.

Scrub – closed shrub vegetation.

Savanna – closed grassland (long or short) with scattered trees.

Steppe – open grassland, herb or dwarf shrub (i.e. bare ground between plants).

Desert – bare ground predominating.

Terms such as 'arboreal' and 'subterranean' are self-explanatory – the latter implies that the animals find most of their food underground and is not used merely to indicate the construction of burrows for shelter. 'Terrestrial' is used to mean living mainly on the ground surface in contrast to 'arboreal' and 'subterranean' as well as to 'aquatic'.

Feeding habits are generally given for higher categories rather than for individual species and then only if they are reasonably constant for the group.

Sources and references

The reference numbers refer to the bibliographies on pages 230 to 241. Since the major sources of information are generally works covering all groups of mammals in a limited geographical region, these are listed in a separate bibliography (p.227) and are not referred to in the text of the list. It can generally be assumed that the principal authority for the recognition of a particular species is the most recent publication in the appropriate geographical bibliography, unless there is in the text of the list a reference to a work dealing with the particular species or higher taxonomic group. Particular attention has been paid to documenting the source for species that are not listed by Honacki *et al.* (p.227), nor in the latest regional works, and to providing references to the original descriptions of genera and species that have been newly described and named during the last five years (as distinct from those that have simply been reinstated as species).

These references are intended primarily to indicate sources where further detail and justification can be found for the acceptance of the species included in the list, e.g. in deciding such issues as whether three named forms should be considered as belonging to one species or to three separate species. They have not necessarily been followed with regard to other aspects such as nomenclature and higher classification. The bibliographies also contain supplementary sources, not referred to in the text, which provide useful additional information but have not been used as authorities for the recognition of species.

Endangered species

An asterisk, *, following the species entry indicates species that are included in the Red Data Books published by the International Union for the Conservation of Nature (IUCN), whether classified as endangered, vulnerable, rare or indeterminate, or in Appendices I or II of the Convention on International Trade in Endangered Species, i.e. species for which trade is restricted or at least monitored on an international basis. In cases where the Convention includes all species of a higher taxon, e.g. all Primates, an asterisk follows the entry for the higher taxon, and for convenience the individual families but not genera nor species are separately marked. An asterisk in parenthesis, (*), indicates that only certain subspecies or populations are included in the above lists, not the entire species.

Domesticated mammals

The naming of domesticated animals and their wild ancestors is confused. On the one hand, both wild and domesticated forms have generally been given separate formal scientific names (e.g. *Canis lupus* for the wolf and *Canis familiaris* for the dog). But it has also been argued that because wild and domestic forms are normally interfertile they should be considered as belonging to the same species. In this case the valid name for the species may be that given originally to the wild form or to the domestic form, depending upon which was the first to be proposed. This leads to many confusing anomalies that conflict with general usage and various proposals have been made to resolve the problem. None has found general acceptance.

The solution followed here is to consider that since clearly distinguishable domesticated forms do not in fact interbreed with the wild species to the extent of losing their separate identity (even although they are potentially capable of doing so), they should not be considered a part of the ancestral wild species and consequently the names applied to domesticated forms should not be used as the names of the wild species. Only the wild species are listed here, but any domesticated derivatives are noted after the range. Names based on domesticated forms are given as synonyms in the first column only when they have frequently been used to include the wild species. An exception is made only in the case of the dromedary which is included in the list since no distinct ancestral species is known. For a more detailed account of the nomenclature of domesticated animals see the appendix in *Evolution of domesticated animals* edited by I.L. Mason (Longman, 1984).

Authorship

In compiling the list the bats (order Chiroptera) have been dealt with by J.E. Hill and the other orders by G.B. Corbet.

Note on the second edition

In bringing the list of species up to date we have not attempted to make any alteration to the higher classification used nor to the sequence followed except to include the pinnipedes in the order Carnivora and to recognize the Megalonychidae, Hylobatidae, Herpestidae and Myocastoridae as families. Changes to the list of species have resulted from the publication of taxonomic revisions and new or revised regional compilations, and from the description of new species, resulting in a net increase of 221 species.

Wherever practicable the spelling of names has been brought into line with the third edition of the *International code of zoological nomenclature* (1985) although some ambiguities remain as to how anomalous original spellings should be treated.

Our grateful thanks are due to all who made constructive comments on the first edition, especially to J. Burton, P. Grubb, K. E. Kinman, J. Nelson and D. W. Yalden. We are also grateful to Mrs S. Lalji for a great deal of secretarial help in the preparation of this edition.

The World List of Mammalian Species

ORDER MONOTREMATA

Monotremes; 3 species; Australia, New Guinea; terrestrial and aquatic predators, mainly on invertebrates.

Family Tachyglossidae

Spiny anteaters (echidnas); 2 species; Australia, New Guinea; terrestrial insectivores.

Tachyglossus

T. aculeatus	Short-nosed echidna	Australia, Tasmania, SE New Guinea; steppe, forest

Zaglossus

Z. bruijnii (*bartoni*)	Long-nosed echidna	New Guinea; forest; *

Family Ornithorhynchidae

One species; a freshwater predator.

Ornithorhynchus

O. anatinus	Platypus	E Australia, Tasmania

ORDER MARSUPIALIA

Marsupials; *c.* 270 species; Australia, New Guinea, (Sulawesi), S America, C America, (N America); all terrestrial habitats; herbivores and predators.

Family Didelphidae

American opossums; *c.* 77 species; S, C America, (N America); mainly forest; predators and omnivores.

Marmosa; (*Thylamys*); mouse-opossums; S, C America; forest, grassland; taxonomy very provisional.

M. aceramarcae	Bolivia
M. agilis	Central S America
M. agricolai	E Brazil
M. alstoni (*cinerea*)	Colombia – Honduras; (in *M. cinerea*?)
M. andersoni	Peru

M. canescens	Greyish mouse-opossum	S Mexico
M. cinerea		N Argentina – Colombia
M. constantiae		C Brazil – N Argentina
M. cracens		Venezuela
M. domina		Amazonian Brazil
M. dryas		W Venezuela, E Colombia
M. elegans		NW Argentina, Bolivia, Peru, Chile
M. emiliae		Para, Brazil
M. formosa		N Argentina
M. fuscata (carri)		Colombia, W Venezuela, Trinidad
M. germana		E Ecuador, Peru
M. grisea		Paraguay
M. handleyi		Colombia; ref. 1.12
M. impavida	Pale mouse-opossum	E Panama – Peru, W Brazil
M. incana		E Brazil
M. invicta	Panama mouse-opossum	Panama
M. karimii		C Brazil
M. lepida		Amazon Basin, Surinam
M. leucastra		N Peru
M. mapiriensis		Bolivia, Peru
M. marica		N Venezuela; montane
M. mexicana	Mexican mouse-opossum	Mexico – Panama
M. microtarsus		SE Brazil, NE Argentina
M. murina		Tropical S America, Tobago
M. noctivaga		Tropical S America
M. ocellata		Bolivia
M. parvidens		Brazil, Peru – Surinam, Colombia
M. phaea		W Colombia, W Ecuador
M. pusilla		Argentina, Bolivia, Paraguay
M. quichua		E Peru
M. rapposa		SE Peru, NE Bolivia
M. regina		Colombia
M. robinsoni (mitis)		Belize – NW South America, Trinidad, Tobago, Grenada
M. rubra		E Ecuador, S Peru
M. scapulata		SE Brazil
M. tatei		Peru

M. tyleriana		Venezuela
M. unduaviensis		W Bolivia
M. velutina		SE Brazil
M. xerophila		Venezuela, Colombia
M. yungasensis		W Bolivia – Ecuador

Monodelphis; short-tailed opossums; S, C America; forest, savanna.

M. adusta	Cloudy short-tailed opossum	Panama – Peru
M. americana	Three-striped short-tailed opossum	Guianas, Brazil, NE Argentina
M. brevicaudata (*touan*)	Seba's short-tailed opossum	Colombia – Surinam – N Argentina; forest
M. dimidiata	Eastern short-tailed opossum	S Brazil, Uruguay, Argentina
M. domestica	Grey short-tailed opossum	E, C Brazil – N Argentina
M. emiliae		Amazon Basin; ref. 1.3
M. henseli	Hensel's short-tailed opossum	S Brazil – N Argentina
M. iheringi		S Brazil
M. kunsi		N Bolivia; lowland forest
M. maraxina		Amazon delta
M. orinoci		Venezuela, etc., savanna
M. osgoodi (*adusta*)	Osgood's short-tailed opossum	W Bolivia, S Peru
M. scalops	Red-headed short-tailed opossum	E Brazil, N Argentina; ref. 1.4
M. sorex	Soricine short-tailed opossum	S Brazil
M. theresa (*americana*)		E Brazil
M. umbristriata		E Brazil
M. unistriata	One-striped short-tailed opossum	SE Brazil

Lestodelphys

L. halli	Patagonian opossum	S Argentina; grassland

Metachirus; (*Philander*)

M. nudicaudatus	Brown four-eyed opossum	Nicaragua – N Argentina

Didelphis; large American opossums; S, C, N America.

D. albiventris (*azarae*)		S America (mainly west)
D. marsupialis (*azarae*)		E Mexico – N Argentina, L Antilles

D. virginiana	Virginian opossum	N Costa Rica – SE Canada

Philander; (*Metachirops*)

P. mcilhennyi		E Peru; dry forest
P. opossum	Grey four-eyed opossum	E Mexico – N Argentina; forest

Lutreolina

L. crassicaudata	Little water opossum	C Argentina – Bolivia – SE Brazil, Guianas, Venezuela; freshwater margins in forest & savanna

Chironectes

C. minimus	Yapok (Water opossum)	S Mexico – N Argentina; freshwater margins

Caluromys; woolly opossums; S, C America; forest.

C. derbianus	S Mexico – Ecuador
C. lanatus	N, C South America
C. philander	N South America – N Argentina

Caluromysiops

C. irrupta	Black-shouldered opossum	SE Peru; forest

Glironia; bushy-tailed opossums; W South America; forest.

G. venusta (*criniger*)	Peru, Ecuador, N Bolivia

Family Microbiotheriidae

One species; (many fossil species)

Dromiciops

D. australis	Colocolos	S, C Chile, W Argentina; forest; arboreal insectivore

Family Caenolestidae

Shrew-opossums (rat-opossums); *c.* 7 species; Andes from Venezuela to Chile; montane forest and grassland; terrestrial insectivores.

Caenolestes

C. caniventer		SW Ecuador

C. convelatus	N Ecuador
C. fuliginosus	Ecuador, N Peru
C. obscurus	Colombia, W Venezuela
C. tatei	SW Ecuador

Lestoros

L. inca	S Peru

Rhyncholestes

R. raphanurus	Chile

Family Dasyuridae

Marsupial mice, marsupial cats, etc.; *c.* 56 species; Australia, New Guinea; desert to forest; carnivores and insectivores.

Murexia; New Guinea; forest.

M. longicaudata	Short-haired marsupial mouse	New Guinea
M. rothschildi	Broad-striped marsupial mouse	E New Guinea

Neophascogale

N. lorentzii	Long-clawed marsupial mouse	New Guinea; montane forest

Phascolosorex; New Guinea; montane.

P. doriae	Red-bellied marsupial mouse	W New Guinea
P. dorsalis	Narrow-striped marsupial mouse	New Guinea

Myoictis

M. melas	Three-striped marsupial mouse	New Guinea; lowland forest

Planigale; planigales; Australia, New Guinea; savanna, steppe.

P. gilesi	Paucident planigale	EC Australia
P. ingrami (*subtilissima*)	Ingram's planigale (Long-tailed planigale)	N Australia
P. maculata	Pygmy planigale (Common planigale)	N, NE Australia
P. novaeguineae	Papua marsupial mouse	E, C New Guinea
P. tenuirostris	Narrow-nosed planigale	E, S Australia

Antechinus; marsupial mice; Australia, New Guinea.

A. bellus	Fawn antechinus	N Australia; savanna
A. flavipes	Yellow-footed antechinus	E, S, SW Australia
A. godmani	Godman's antechinus	NE Queensland; forest
A. leo	Cinnamon antechinus	N Queensland; ref. 1.13
A. melanurus	Black-tailed antechinus	New Guinea; forest
A. minimus	Swamp antechinus	SE Australia, Tasmania; grassland
A. naso	Long-nosed antechinus	W New Guinea; montane
A. stuartii	Brown antechinus	E Australia; forest, woodland
A. swainsonii	Dusky antechinus	E Australia, Tasmania; forest
A. wilhelmina	Lesser antechinus	New Guinea; montane

Parantechinus; (*Antechinus*); Australia; ref. 1.16.

P. apicalis	Dibbler	SW Australia; scrub; *
P. bilarni	Harney's marsupial mouse	N Australia

Pseudantechinus; (*Antechinus*); Australia; ref. 1.16.

P. macdonnellensis	Fat-tailed marsupial mouse (Red-eared antechinus)	W, N Australia; rocky hills

Dasykaluta; (*Antechinus*); Australia; ref. 1.16.

D. rosamondae	Little red marsupial mouse	NW Australia; grassland

Phascogale

P. calura	Red-tailed wambenger	SW, († C, SE) Australia; dry woodland; *
P. tapoatafa	Tuan (Common wambenger)	Australia

Dasycercus

D. cristicauda	Mulgara	C Australia; desert, grassland

Dasyurus; (*Satanellus*); marsupial cats; Australia, New Guinea; forest, woodland.

D. albopunctatus	New Guinea marsupial cat	New Guinea; forest
D. geoffroii	Chuditch	W († N, E) Australia, C New Guinea; woodland
D. hallucatus	Satanellus	NE, N, NW Australia
D. maculatus	Tiger cat	E, SE Australia, Tasmania
D. viverrinus	Quoll	SE Australia?, Tasmania

Sarcophilus
S. harrisii   Tasmanian devil   Tasmania; dry forest

Dasyuroides
D. byrnei   Kowari   C Australia; desert,
                                                     grassland

Ningaui; Australia; desert; ref. 1.5.
N. ridei   Wongai ningaui   W, C Australia
N. timealeyi   Pilbara ningaui   NW Australia
N. yvonneae   SW, S, SE Australia

Sminthopsis; (Antechinomys); dunnarts; Australia, S New Guinea.
S. aitkeni   S Australia; ref. 1.6
  (murina)
S. butleri   Carpentarian dunnart   N Australia
S. crassicaudata   Fat-tailed dunnart   Australia; grassland,
                                                                        woodland
S. dolichura   SW, S Australia; ref. 1.6
  (murina)
S. douglasi   Julia Creek dunnart   C Queensland; *
S. gilberti   SW Australia; ref. 1.6
  (murina)
S. granulipes   White-tailed dunnart   SW Australia; woodland,
                                                                            scrub
S. griseoventer   SW Australia; ref. 1.6
  (murina)
S. hirtipes   Hairy-footed dunnart   W, C Australia; desert
S. laniger   Kultarr   E, W, C Australia;
  (spenceri)   (Wuhl-wuhl)   savanna, steppe
S. leucopus   White-footed dunnart   SE Australia, Tasmania;
                                                                            forest
S. longicaudata   Long-tailed dunnart   W Australia; *
S. macroura   Stripe-faced dunnart   E Australia
                             (Darling Downs
                             dunnart)
S. murina   Common dunnart   E Australia; forest; ref. 1.6
S. ooldea   Ooldea dunnart   S, C Australia
S. psammophila   Sandhill dunnart   C, S Australia; desert
                                                                        grassland; *
S. virginiae   Red-cheeked dunnart   N Australia, S New
  (rufigenis)   Guinea; forest, scrub
S. youngsoni   Lesser hairy-footed dunnart   NW Australia; desert; ref.
                                                                                1.7

Family Myrmecobiidae

One species.

Myrmecobius
M. fasciatus	Numbat	SW († S, SE) Australia;
	(Banded anteater)	dry forest, woodland; *

Family Thylacinidae

One species.

Thylacinus
T. cynocephalus †	Thylacine	Tasmania; woodland;
	(Tasmanian wolf)	? extinct; *

Family Notoryctidae

One species.

Notoryctes
N. typhlops	Marsupial mole	S, W Australia; desert

Family Peramelidae

Bandicoots; 21 species; Australia, New Guinea; forest, (grassland).

Peroryctes; New Guinea bandicoots; New Guinea.
P. broadbentii	Giant bandicoot	SE New Guinea; (in *P. raffrayanus?*)
P. longicauda	Striped bandicoot	New Guinea
P. papuensis	Papuan bandicoot	SE New Guinea
P. raffrayanus	Raffray's bandicoot	New Guinea

Microperoryctes
M. murina	Mouse-bandicoot	W New Guinea

Perameles; long-nosed bandicoots; Australia.
P. bougainville (*fasciata*)	Marl (Barred bandicoot)	Bernier & Dorre Is, W Australia, † mainland Australia; *
P. eremiana †	Desert bandicoot	C Australia; desert grassland; ? extinct; *
P. gunnii	Gunn's bandicoot (Eastern barred bandicoot)	SE Australia, Tasmania; woodland
P. nasuta	Long-nosed bandicoot	E Australia; forest, woodland

Echymipera; spiny bandicoots; New Guinea, Australia.

E. clara	White-lipped bandicoot	W New Guinea; *
E. kalubu	Spiny bandicoot	New Guinea
E. rufescens	Rufescent bandicoot	New Guinea,
		N Queensland; forest

Rhynchomeles

R. prattorum	Ceram bandicoot	Ceram Island

Isoodon; short-nosed bandicoots; Australia, S New Guinea; ref. 1.8.

I. arnhemensis		N Australia
I. auratus	Golden bandicoot	W, N, († C) Australia
I. barrowensis		Barrow I, W Australia
I. macrourus	Brindled bandicoot	NE, N Australia,
	(Northern brown	S New Guinea
	bandicoot)	
I. nauticus		Nuyts Arch., S Australia
I. obesulus	Southern brown bandicoot	SE, SW Australia,
		Tasmania
I. peninsulae		N Queensland

Chaeropus

C. ecaudatus †	Pig footed bandicoot	SW, S Australia;
		woodland, scrub,
		grassland; ? extinct; *

Family Thylacomyidae

Rabbit-bandicoots; 2 species; Australia; *.

Macrotis

M. lagotis	Common rabbit-bandicoot	C, NW Australia;
	(Bilby)	woodland, savanna
M. leucura †	Lesser rabbit-bandicoot	C Australia; sandhills;
	(Lesser bilby)	? extinct

Family Phalangeridae

Phalangers; 14 species; Australia, New Guinea, (Sulawesi); forest.

Trichosurus; Australia; forest, woodland.

T. arnhemensis	Northern brush possum	NW Australia
T. caninus	Bobuck	E Australia; montane
	(Mountain possum)	forest
T. vulpecula	Brush-tailed possum	Australia, Tasmania, [New
		Zealand]

Wyulda

W. squamicaudata	Scaly-tailed possum	NW Australia; woodland

Phalanger; cuscuses; N Australia; New Guinea, Sulawesi; forest, scrub.

P. carmelitae (orientalis)		New Guinea; montane; *
P. celebensis		Sulawesi, Molucca Is, etc.
P. gymnotis (leucippus)	Ground cuscus (Grey phalanger)	E New Guinea, Aru Islands
P. interpositus (orientalis)	Stein's cuscus	New Guinea; *
P. lullulae (orientalis)	Woodlark cuscus	Woodlark I, New Guinea; *
P. maculatus	Spotted cuscus (Spotted phalanger)	N Queensland, New Guinea; *
P. orientalis	Grey cuscus (Common phalanger)	N Queensland, New Guinea, Ceram, Solomon Is; *
P. rufoniger (atrimaculatus)	Black-spotted cuscus	N New Guinea; *
P. ursinus	Bear cuscus	Sulawesi, etc.
P. vestitus	Silky cuscus	New Guinea

Family Burramyidae

Pygmy possums; 7 species; Australia, New Guinea; forest; insectivores, nectar-feeders.

Cercartetus; (Eudromicia); dormouse-possums.

C. caudatus	Long-tailed pygmy possum	NE Queensland, New Guinea; forest
C. concinnus	Southwestern pygmy possum	SW, S, SE Australia; dry forest
C. lepidus	Tasmanian pygmy possum (Little pygmy possum)	Tasmania, Victoria, Kangaroo I, (S Australia); dry forest
C. nanus	Eastern pygmy possum	SE Australia, Tasmania; forest

Distoechurus

D. pennatus	Feather-tailed possum	New Guinea; montane forest

Acrobates

A. pygmaeus	Pygmy glider (Flying mouse) (Feather-tail glider)	E, SE Australia; dry forest

Burramys

B. parvus	Mountain pygmy possum	Victoria, New South Wales; *

Family Petauridae

Ringtails, ringtail possums, gliding phalangers; c. 23 species; Australia, New Guinea; forest.

Gymnobelideus

G. leadbeateri	Leadbeater's possum	Victoria; wet forest; *

Petaurus

P. abidi	Northern glider	NC New Guinea; ref. 1.14
P. australis	Fluffy glider (Yellow-bellied glider)	E Australia; coastal montane forest
P. breviceps	Sugar glider	E, N Australia, New Guinea; [Tasmania]
P. norfolcensis	Squirrel glider	E Australia

Pseudocheirus; (Hemibelideus, Petropseudes); ringtails; Australia, New Guinea; forest.

P. albertisii	D'Albertis' ringtail	W New Guinea
P. archeri	Green ringtail	NE Queensland; rain forest
P. canescens	Lowland ringtail	New Guinea
P. caroli	Weyland ringtail	W New Guinea
P. corinnae	Eastern ringtail	New Guinea; montane forest
P. cupreus	Coppery ringtail	New Guinea; montane forest
P. dahli	Rock-haunting ringtail	N, NW Australia; rocks in savanna
P. forbesi	Moss-forest ringtail	New Guinea; montane forest
P. herbertensis	Mongan (Herbert River ringtail)	NE Queensland; rain forest
P. lemuroides	Brushy-tailed ringtail (Lemuroid ringtail)	NE Queensland; rain forest
P. mayeri	Pygmy ringtail	New Guinea; montane forest
P. peregrinus	Common ringtail	E, SW Australia, Tasmania; forest
P. schlegeli	Arfak ringtail	W New Guinea

Petauroides; (*Schoinobates*); ref. 1.9.

P. *volans*	Greater glider	E Australia; forest

Dactylopsila; (*Dactylonax*); striped possums; New Guinea, (Australia); forest.

D. *megalura*	Large-tailed possum	W New Guinea
D. *palpator*	Long-fingered possum	E New Guinea
D. *tatei*	Fergusson striped possum	Fergusson I (E of New Guinea)
D. *trivirgata*	Common striped possum	New Guinea, NE Queensland

Family Macropodidae

Kangaroos, wallabies; *c.* 54 species; Australia, New Guinea; steppe, savanna, forest; grazers, browsers.

Hypsiprymnodon

H. *moschatus*	Musk rat-kangaroo	NE Queensland; forest

Potorous; potoroos; Australia; grass, scrub.

P. *longipes*	Long-footed potoroo	E Victoria; ref. 1.15; *
P. *platyops* †	Broad-faced potoroo	SW Australia; extinct
P. *tridactylus* (*apicalis*)	Long-nosed potoroo	E, SE, SW Australia, Tasmania

Bettongia; Australia; steppe, dry woodland; *

B. *gaimardi* (*cuniculus*)	Eastern bettong (Tasmanian bettong)	Tasmania, † E Australia
B. *lesueuri*	Boodie (Burrowing bettong)	W Australia (small islands only), († W, C Australia)
B. *penicillata* (*tropica*)	Woylie (Brush-tailed bettong)	SW, († S) Australia, E Queensland; *

Aepyprymnus

A. *rufescens*	Rufous rat-kangaroo (Rufous bettong)	E Australia; forest, woodland

Caloprymnus

C. *campestris* †	Desert rat-kangaroo	C Australia; desert; ? extinct; *

Thylogale; pademelons, scrub wallabies; E Australia, New Guinea.

T. *billardierii*	Red-bellied pademelon (Tasmanian wallaby)	Tasmania, † SE Australia; scrub
T. *brunii*	Dusky wallaby	New Guinea, Bismarck Is.

T. stigmatica	Red-legged pademelon	E Australia, SE New Guinea; forest
T. thetis	Red-necked pademelon	E Australia; forest

Petrogale; rock wallabies; Australia; rocks in forest and grassland.

P. brachyotis	Short-eared rock wallaby	N Australia
P. burbidgei	Warabi	W Australia
P. penicillata (*godmani*) (*inornata*) (*lateralis*)	Brush-tailed rock wallaby (Western rock wallaby)	Australia, [Hawaii]
P. persephone	Prosperine rock wallaby	NE Queensland; ref. 1.10
P. purpureicollis	Purple-necked rock wallaby	W Queensland
P. rothschildi	Rothschild's rock wallaby	W Australia
P. xanthopus	Yellow-footed rock wallaby (Ring-tailed rock wallaby)	E, SE Australia

Peradorcas; (*Petrogale*).

P. concinna	Little rock wallaby (Nabarlek)	N Australia

Lagorchestes; hare-wallabies; Australia; grassland.

L. conspicillatus	Spectacled hare-wallaby	N Australia
L. hirsutus	Western hare-wallaby (Rufous hare-wallaby)	W, C Australia (islands only in W); *
L. leporides †	Eastern hare-wallaby	SE Australia; extinct

Setonyx

S. brachyurus	Quokka (Short-tailed wallaby)	SW Australia; scrub

Lagostrophus

L. fasciatus	Banded hare-wallaby	Bernier & Dorre Is, W Australia, † W, S Australia; *

Macropus; (*Megaleia, Protemnodon*); kangaroos, wallabies; Australia, New Guinea.

M. agilis	River wallaby (Agile wallaby) (Sand wallaby)	N Australia, New Guinea; savanna
M. antelopinus	Antelope-kangaroo	N Australia; savanna
M. bernardus	Bernard's wallaroo (Black wallaroo)	N Australia; rocky savanna; *

M. dorsalis	Black-striped wallaby	E Australia; forest
M. eugenii	Tammar wallaby (Dama wallaby)	SW, S Australia; dry forest
M. fuliginosus	Western grey kangaroo	SW, S Australia; woodland
M. giganteus	Easte n grey kangaroo	E Australia, Tasmania; woodland
M. greyi †	Toolache	S Australia, ? extinct; (in *M. irma*?)
M. irma	Western brush wallaby	SW Australia; scrub, dry forest
M. parma	Parma wallaby	SE Australia; forest, scrub; *
M. parryi	Whiptail (Pretty-face wallaby)	NE Australia; woodland
M. robustus	Wallaroo (Hill kangaroo)	Australia; rocky hills in forest, grassland & desert
M. rufogriseus	Red-necked wallaby (Bennett's wallaby)	E, SE Australia, Tasmania [England, New Zealand]; scrub, woodland
M. rufus	Red kangaroo	Australia; grassland

Wallabia

W. bicolor	Swamp wallaby	E Australia; forest, thicket

Onychogalea; nail-tailed wallabies; Australia; steppe, savanna.

O. fraenata	Bridle nail-tailed wallaby	S Queensland, († SE Australia); *
O. lunata	Crescent nail-tailed wallaby	C, († SW) Australia; ? extinct; *
O. unguifera	Northern nail-tailed wallaby	N Australia

Dendrolagus; tree kangaroos; New Guinea, Queensland; forest; ref. 1.11.

D. bennettianus	Bennett's tree kangaroo	NE Queensland; *
D. dorianus	Unicolored tree kangaroo	New Guinea; (*)
D. inustus	Grizzled tree kangaroo	W, N New Guinea; *
D. lumholtzi	Lumholtz's tree kangaroo	NE Queensland; *
D. matschiei (*deltae*) (*spadix*) (*goodfellowi*)	Matschie's tree kangaroo	New Guinea; (*)
D. ursinus	Vogelkop tree kangaroo	NW New Guinea; *

Dorcopsis; (*Dorcopsulus*); forest wallabies; New Guinea; forest.

D. atrata	Black forest wallaby	Goodenough I, E New Guinea; *
D. hageni	Greater forest wallaby	N New Guinea
D. macleayi	Papuan forest wallaby	E New Guinea; *
D. muelleri (*veterum*)	Common forest wallaby	New Guinea
D. vanheurni	Lesser forest wallaby	New Guinea; (in *D. macleayi* ?)

Family Phascolarctidae

One species.

Phascolarctos

P. cinereus	Koala	E Australia; dry forest

Family Vombatidae

Wombats; 3 species; E, S Australia, Tasmania.

Vombatus

V. ursinus	Common wombat	E Australia, Tasmania; forest, scrub

Lasiorhinus

L. krefftii (*barnardi*) (*gillespiei*)	Queensland hairy-nosed wombat	S Queensland, († SE Australia); *
L. latifrons	Southern hairy-nosed wombat	S Australia

Family Tarsipedidae

One species.

Tarsipes

T. spenserae (*rostratus*)	Honey possum	SW Australia

ORDER EDENTATA

Edentates; *c.* 29 species; S, C, (N) America; forest – grassland; predators on invertebrates, (herbivores); ref. 2.1.

Family Myrmecophagidae

American anteaters; 4 species; S, C America; forest, savanna; terrestrial and arboreal ant-eaters.

Myrmecophaga
M. tridactyla	Giant anteater	N Argentina – Guatemala; forest, savanna; *

Tamandua; collared anteaters; S, C America.
T. mexicana	Northern tamandua	S Mexico – Peru, Venezuela
T. tetradactyla (*longicaudata*)	Southern tamandua	Venezuela – Uruguay; (*)

Cyclopes
C. didactylus	Pygmy anteater (Silky anteater)	S Mexico – Bolivia, Brazil; forest

Family Bradypodidae

Three-toed sloths; 3 species; S, C America; forest; arboreal herbivores.

Bradypus; (*Scaeopus*); three-toed sloths.
B. torquatus	Maned sloth	SE Brazil; *
B. tridactylus	Pale-throated sloth	Guianas, etc.
B. variegatus (*boliviensis*) (*griseus*) (*infuscatus*)	Brown-throated sloth (Grey three-toed sloth)	Honduras – N Argentina; (*)

Family Megalonychidae

Two-toed sloths; 2 species; tropical S, C America; arboreal herbivores; includes also the extinct ground sloths.

Choloepus; two-toed sloths; formerly included in Bradypodidae or Choloepidae.
C. didactylus	Linné's two-toed sloth	Amazon Basin – Venezuela, Guianas
C. hoffmanni	Hoffmann's two-toed sloth	Nicaragua – C Brazil; (*)

Family Dasypodidae

Armadillos; *c.* 20 species; S, (C, N) America; grassland, savanna, (forest); terrestrial; mainly predators on invertebrates.

Chaetophractus; (*Euphractus*); hairy armadillos, etc.
C. nationi	Andean hairy armadillo	W Bolivia, NW Argentina
C. vellerosus	Screaming armadillo	Bolivia, Argentina

Euphractus; (*Chaetophractus*)

E. sexcinctus	Yellow armadillo (Six-banded armadillo)	Uruguay – Surinam, E of Andes
E. villosus	Larger hairy armadillo	C Argentina, Uruguay, Paraguay

Zaedyus

Z. pichiy	Pichi	S Argentina, Chile; grassland

Priodontes

P. maximus (*giganteus*)	Giant armadillo	Venezuela – N Argentina; *

Cabassous; naked-tailed armadillos; S, (C) America.

C. centralis	Northern naked-tailed armadillo	Honduras – W Venezuela
C. chacoensis	Chacoan naked-tailed armadillo	W Paraguay, etc.; ref. 2.2
C. tatouay	Greater naked-tailed armadillo	N Argentina – S Brazil, Uruguay; (*)
C. unicinctus	Southern naked-tailed armadillo	Guyana – N Argentina

Tolypeutes; three-banded armadillos.

T. matacus	Southern 3-banded armadillo	Bolivia – Argentina
T. tricinctus	Brazilian 3-banded armadillo	NE, C Brazil; *

Dasypus; long-nosed armadillos.

D. hybridus	Southern long-nosed armadillo	C Argentina, Uruguay – S Brazil
D. kappleri	Greater long-nosed armadillo	Peru – Surinam
D. novemcinctus	Common long-nosed armadillo (Nine-banded armadillo)	Uruguay – SE USA
D. pilosus	Hairy long-nosed armadillo	Peru; montane
D. sabanicola	Northern long-nosed armadillo	Venezuela, E Colombia
D. septemcinctus	Brazilian long-nosed armadillo	E, C Brazil

Chlamyphorus; (*Calyptophractus, Burmeisteria*); fairy armadillos; *.

C. *retusus*	Chacoan fairy armadillo (Greater fairy armadillo) (Burmeister's armadillo)	Bolivia – N Argentina; grassland
C. *truncatus*	Lesser fairy armadillo (Pink fairy armadillo)	CW Argentina; steppe

ORDER INSECTIVORA

Insectivores; *c*. 354 species; Eurasia, Africa, Madagascar, N, C, (S) America; terrestrial, subterranean; predators on insects and other invertebrates.

Family Solenodontidae

Solenodons; 2 species; Cuba, Hispaniola; terrestrial omnivores.

Solenodon; (*Atopogale*).

S. *cubanus*	Cuban solenodon	Cuba; *
S. *paradoxus*	Haitian solenodon	Hispaniola; *

Family Tenrecidae

Tenrecs, otter-shrews; *c*. 32 species; Madagascar, W, C Africa; terrestrial, (freshwater) predators on invertebrates.

Tenrec; (*Centetes*).

T. *ecaudatus*	Tail-less tenrec	Madagascar, Comoro Is; [Seychelles, Mauritius, Reunion]; dry forest

Setifer; (*Ericulus, Dasogale*); ref. 3.1.

S. *setosus* (*fontoynonti*)	Greater hedgehog-tenrec	Madagascar; forest, scrub

Hemicentetes

H. *semispinosus* (*nigriceps*)	Streaked tenrec	Madagascar; forest, scrub

Echinops

E. *telfairi*	Lesser hedgehog-tenrec	SW Madagascar; dry forest

Oryzorictes; (*Nesoryctes*); mole-tenrecs (rice tenrecs); Madagascar; wet forest, fields.

O. *hova*		C Madagascar
O. *talpoides*		NW Madagascar
O. *tetradactylus*		C Madagascar

Microgale; (*Leptogale, Nesogale, Paramicrogale*); shrew-tenrecs; Madagascar; forest; taxonomy very provisional.

M. brevicaudata	NE Madagascar
M. cowani	E Madagascar
M. crassipes	C Madagascar
M. decaryi	S Madagascar
M. dobsoni	N Madagascar
M. drouhardi	N, SE Madagascar
M. gracilis	S Madagascar; (in *Leptogale?*)
M. longicaudata	E Madagascar
M. longirostris	SE Madagascar
M. majori	E Madagascar
M. melanorrhachis	E Madagascar
M. occidentalis	W Madagascar
M. parvula	N Madagascar
M. principula	SE Madagascar
M. prolixicaudata	N Madagascar
M. pusilla	SE Madagascar
M. sorella	N Madagascar
M. taiva	SE Madagascar
M. talazaci	E Madagascar
M. thomasi	E Madagascar

Limnogale

L. mergulus	Web-footed tenrec	Madagascar; streams

Geogale; (*Cryptogale*).

G. aurita		NE, SW Madagascar

Potamogale

P. velox	Giant otter-shrew	C Africa; freshwater in forest

Micropotamogale

M. lamottei	Nimba otter-shrew	Guinea, etc.; montane streams
M. ruwenzorii	Ruwenzori otter-shrew	Ruwenzori – L Kivu; montane streams

Family Chrysochloridae

Golden moles; *c.* 17 species; S Africa to Cameroun and Somalia; desert to forest; subterranean predators; taxonomy very provisional, especially at generic level.

Chrysochloris

C. asiatica (*? visagiei*)	Cape golden mole	W Cape Province
C. stuhlmanni (*fosteri*)	Stuhlmann's golden mole	E Africa

Eremitalpa

E. granti	Grant's golden mole	SW Africa; coastal desert

Calcochloris; (*Amblysomus*).

C. obtusirostris	Yellow golden mole	S Africa

Cryptochloris

C. wintoni	De Winton's golden mole	W Cape Province; desert
C. zyli	Van Zyl's golden mole	SW Cape Province

Amblysomus

A. gunningi	Gunning's golden mole	Transvaal; montane forest
A. hottentotus	Hottentot golden mole	S Africa
A. iris	Zulu golden mole	SE Africa
A. julianae	Juliana's golden mole	Transvaal; *

Chlorotalpa

C. arendsi	Arends' golden mole	E Zimbabwe; montane
C. duthiae	Duthie's golden mole	S Cape Province
C. leucorhina	Congo golden mole	WC Africa
C. sclateri	Sclater's golden mole	S Africa
C. tytonis	Somali golden mole	Somalia

Chrysospalax

C. trevelyani	Giant golden mole	SE Cape Province; forest; *
C. villosus	Rough-haired golden mole	SE Africa; grassland

Family Erinaceidae

Hedgehogs, moonrats; *c.* 19 species; Eurasia (except boreal zones), Africa; forest to desert; terrestrial predators on invertebrates.

Subfamily Galericinae (Echinosoricinae)

Moonrats, gymnures; 6 species; SE Asia; forest.

Echinosorex; (*Gymnurus*).

E. gymnurus	Moonrat	Malaya, Sumatra, Borneo; forest

Hylomys; (*Neotetracus, Neohylomys*); ref. 3.2.

H. hainanensis	Hainan moonrat	Hainan I, China
H. sinensis	Shrew-hedgehog	S China, Indochina; forest
H. suillus	Lesser moonrat	Burma – Malaya, Sumatra, Java, Borneo; forest

Podogymnura

P. aureospinula	Spiny moonrat	Dinagat I, Philippines; ref. 3.3
P. truei	Mindanao moonrat	Mindanao, Philippines; forest; *

Subfamily Erinaceinae

Spiny hedgehogs; 13 species; Europe, Asia (temperate and SW), Africa.

Erinaceus; woodland hedgehogs; Europe, N Asia.

E. amurensis	Manchurian hedgehog	Manchuria, etc.; forest, grassland
E. concolor	E European hedgehog	SE Europe – Syria
E. europaeus	W European hedgehog	W, N Europe, [New Zealand]; forest, grassland

Atelerix; (*Erinaceus, Aethechinus*); African hedgehogs; Africa; mainly savanna.

A. albiventris	Four-toed hedgehog	Senegal – Zambezi; savanna
A. algirus	Algerian hedgehog	NW Africa, [SW Europe]; scrub, steppe
A. frontalis	S African hedgehog	S Africa – Zambezi; *
A. sclateri	Somali hedgehog	N Somalia

Hemiechinus; steppe hedgehogs; W, C Asia.

H. auritus	Long-eared hedgehog	Libya – Mongolia & NW India
H. dauuricus	Daurian hedgehog	E of Gobi Desert; dry steppe
H. hughi (*sylvaticus*)		Shanxi, China; (in *H. dauuricus*?)

Paraechinus; desert hedgehogs; Sahara to India.

P. aethiopicus (*deserti*) (*dorsalis*)	Desert hedgehog	Sahara, Arabia; desert
P. hypomelas	Brandt's hedgehog	Iran, Turkestan – Pakistan; desert, dry steppe

P. micropus   Indian hedgehog   NW, SW India
  (*nudiventris*)

Family Soricidae

Shrews; *c.* 254 species; Eurasia, Africa, N America to northern S America; forest to desert; terrestrial (or partially aquatic) insectivores.

Sorex; (*Microsorex*); red-toothed shrews; N America, N Eurasia; tundra, forest; ref. 3.4 (N America).

S. alaskanus	Glacier Bay water shrew	S Alaska; (in *S. palustris*?)
S. alpinus	Alpine shrew	Europe; montane forest
S. araneus	Eurasian common shrew	Europe – R Yenesei; forest, tundra
S. arcticus	Arctic shrew (Black-backed shrew)	Alaska, Canada, N USA; wet forest
S. arizonae	Arizona shrew	SE Arizona, etc.; montane; close to *S. ventralis*
S. asper	Tien Shan shrew	C Asia; montane
S. bedfordiae	Lesser striped shrew	Gansu – Himalayas; montane forest
S. bendirii	Pacific water shrew	W coast USA; wet forest
S. buchariensis (*thibetanus*)	Pamir shrew	Tibet etc.
S. caecutiens	Laxmann's shrew	E Europe – Japan; con. forest, tundra
S. caucasicus	Caucasian shrew	Caucasus, etc.
S. cinereus	Masked shrew (American common shrew)	N USA, Canada, NE Siberia; forest, tundra
S. coronatus		France, etc.; close to *S. araneus*
S. cylindricauda	Greater striped shrew	Sichuan, China; montane forest
S. daphaenodon	Large-toothed Siberian shrew	Siberia, NE China; con. forest, tundra
S. dispar	Long-tailed shrew	NE USA
S. emarginatus		Mexico; close to *S. ventralis*
S. fontinalis		E USA
S. fumeus	Smoky shrew	SE Canada, NE USA
S. gaspensis	Gaspé shrew	Gaspé Peninsula, Quebec; close to *S. dispar*
S. gracillimus	Slender shrew	NE Asia, N Japan
S. granarius		Spain; close to *S. araneus*
S. haydeni	Prairie shrew	N USA, S Canada; close to *S. cinereus*

S. hosonoi	Azumi shrew	C Honshu, Japan; montane
S. hoyi (thompsoni)	American pygmy shrew	Canada, NE USA; con. forest; (formerly in Microsorex)
S. hydrodromus	Pribilof shrew	St Paul I, Alaska
S. jacksoni (? pribilofensis) (ugyunak)	Barrenground shrew	Alaska;? N Canada, NE Siberia
S. longirostris	Southeastern shrew	SE USA
S. lyelli	Mount Lyell shrew	California; montane
S. macrodon	Large-toothed shrew	Veracruz, Mexico; forest
S. merriami	Merriam's shrew	W USA; high steppe
S. milleri	Carmen Mountain shrew	NE Mexico; montane
S. minutissimus	Least shrew	Scandinavia, Siberia, S China, Japan; con. forest
S. minutus	Eurasian pygmy shrew	Europe – C Siberia; forest, tundra
S. mirabilis	Giant shrew	E Siberia, N Korea
S. monticolus (obscurus)	Dusky shrew	Alaska, W Canada, W USA, NW Mexico; tundra, forest
S. nanus	Dwarf shrew	Wyoming – New Mexico; montane
S. oreopolis	Mexican long-tailed shrew	Mexico; montane
S. ornatus (juncensis) (sinuosus)	Ornate shrew	California, NW Mexico
S. pacificus	Pacific shrew	W USA; coastal forest
S. palustris	American water shrew	Canada, USA; wet forest
S. planiceps (minutus)	Kashmir shrew	W Himalayas; ? W China
S. preblei	Preble's shrew	E Oregon – Montana, USA
S. raddei	Radde's shrew	Caucasus, etc.
S. roboratus (vir)	Flat-skulled shrew	E Siberia – Altai
S. samniticus (araneus)	Apennine shrew	S, C Italy
S. saussurei	Saussure's shrew	C Mexico – Guatemala; montane
S. sclateri	Sclater's shrew	Chiapas, SE Mexico
S. sinalis (isodon)		N Europe – N China; con. forest

S. stizodon	San Cristobal shrew	Chiapas, SE Mexico
S. tenellus	Inyo shrew	California, Nevada
S. trowbridgii	Trowbridge shrew	W coast USA; forest
S. tundrensis (*arcticus*)	Tundra shrew	Alaska, NE Siberia
S. unguiculatus	Long-clawed shrew	E Siberia, N Japan
S. vagrans (*trigonirostris*)	Vagrant shrew (Wandering shrew)	W USA, S Mexico; marsh, wet forest
S. veraepacis	Verapaz shrew	SE Mexico; montane
S. ventralis		Mexico
S. volnuchini (*minutus*)	Caucasian pygmy shrew	Caucasus

Soriculus; (*Chodsigoa, Episoriculus*); mountain shrews; Himalayas, China; montane forest; ref. 3.5.

S. caudatus	Hodgson's brown-toothed shrew	Himalayas – S China, Burma
S. fumidus		Taiwan; close to *S. caudatus*
S. hypsibius	De Winton's shrew	China
S. lamula (*hypsibius*)		Yunnan – Gansu
S. leucops (*baileyi*) (*gruberi*)	Indian long-tailed shrew	C Nepal – S China – N Vietnam
S. macrurus (*leucops*)		C Nepal – Vietnam
S. nigrescens	Himalayan shrew	Himalayas – N Burma
S. parca (*smithii*) (*lowei*)		Thailand – Sichuan
S. salenskii	Salenski's shrew	N Sichuan
S. smithii	Smith's shrew	Sichuan – Shaanxi

Neomys; Eurasian water shrews; N Eurasia; wet forest, streams, marshes.

N. anomalus	Southern water shrew	S, E Europe
N. fodiens	Eurasian water shrew	W Europe – E Siberia, Korea
N. schelkovnikovi	Transcaucasian water shrew	Armenia, Georgia

Blarina; American short-tailed shrews; E N America; forest, grassland.

B. brevicauda	Northern short-tailed shrew	E USA, SE Canada
B. carolinensis (*telmalestes*)	Southern short-tailed shrew	SE USA

B. hylophaga   Elliot's short-tailed shrew   S, SE USA
  (*carolinensis*)

Blarinella

B. quadraticauda   Chinese short-tailed shrew   S China; montane forest

Cryptotis; small-eared shrews; E USA – Ecuador, Surinam; forest, (grassland).
C. avia   Colombia
C. endersi   Ender's small-eared shrew   W Panama
C. goldmani   Goldman's small-eared   S Mexico; wet montane
                                  shrew   forest
C. goodwini   Goodwin's small-eared   S Guatemala, S Mexico;
                                  shrew   montane forest
C. gracilis   Talamancan small-eared   Honduras – Panama; forest
                                  shrew
C. magna   Big small-eared shrew   Oaxaca, S Mexico;
                                                                   montane forest
C. mexicana   Mexican small-eared shrew   S, E Mexico; wet montane
  (*phillipsi*)   forest
C. montivaga   S Ecuador
C. nigrescens   Blackish small-eared shrew   S Mexico – Panama; scrub,
                                                                   forest
C. parva   American least shrew   C, E USA – Panama;
                                                                   grassland, scrub, forest
C. squamipes   W Colombia
C. thomasi   Ecuador – Venezuela

Notiosorex

N. crawfordi   Desert shrew   SW USA, Mexico;
                                  (Grey shrew)   semidesert scrub

Megasorex; (*Notiosorex*).

M. gigas   Giant Mexican shrew   SW Mexico
                                  (Merriam's desert shrew)

Crocidura; (*Praesorex*); white-toothed shrews; Eurasia, Africa; forest to semi-
desert; taxonomy very provisional in some areas; ref. 3.6 (Oriental Region).
C. allex   Kenya, N Tanzania
C. andamanensis   S Andaman I, Indian
                                                                   Ocean
C. attenuata   Himalayas – S China –
  (*aequicauda*)   Java, Taiwan
  (*trichura*)
C. baileyi   Ethiopia; montane

C. beatus		Mindanao, Philippines
C. beccarii		Sumatra; doubtful species
C. bloyeti		C Tanzania
C. bottegi		W Africa, Ethiopia, N Kenya
C. bovei		Zaire
C. butleri		Sudan, N Kenya, S Somalia
C. buettikoferi		W Africa; forest
C. caliginea		NE Zaire; forest
C. cinderella		Gambia, ? Mali
C. congobelgica		NE Zaire
C. crenata		Gabon
C. crossei		Nigeria – Ivory Coast; forest
C. cyanea	Reddish-grey musk shrew	Ethiopia – S Africa
C. denti		C Africa
C. dolichura		Guinea – Uganda
C. douceti		Ivory Coast, Guinea
C. dsinezumi		Japan except Hokkaido, ? Taiwan
C. edwardsiana		Jolo I, Philippines
C. eisentrauti		Mt Cameroun
C. elgonius		W Kenya, N Tanzania
C. elongata		Sulawesi
C. erica		W Angola
C. fischeri		S Ethiopia – Zaire
C. flavescens (*occidentalis*) (*olivieri*) (*ferruginea*)	Greater musk shrew	Africa except NW; savanna, forest
C. floweri	Flower's shrew	Egypt
C. foucauldi		Morocco; (in *C. russula*?)
C. foxi		Nigeria – Ghana; Guinea savanna
C. fuliginosa (*malayana*)		S China – Malaya, Sumatra, Java, Borneo, Timor
C. fulvastra (*arethusa*) (*sericea*)		Mali – Ethiopia, Kenya; ref. 3.8
C. fumosa		E Africa; montane forest
C. fuscomurina (*bicolor*)	Tiny musk shrew	Senegal – Ethiopia – N Cape Prov., Namibia; savanna; ref. 3.9

C. glassi		Harar, Ethiopia; montane
C. gracilipes (hildegardeae)	Peters' musk shrew	W, E S Africa
C. grandiceps		Ivory Coast – S Nigeria; ref. 3.11
C. grandis		Mindanao, Philippines
C. grassei		C Africa
C. grayi		Luzon, Philippines
C. greenwoodi		S Somalia
C. gueldenstaedtii (? cypria)		Caucasus – Israel, Crete, ? Cyprus
C. halconis		Mindoro, Philippines
C. hirta	Lesser red musk shrew	S Africa – S Somalia
C. hispida	Andaman spiny shrew	Middle Andaman I, Indian Ocean
C. horsfieldii		Sri Lanka, Indochina, Hainan, Taiwan, Ryukyu Is
C. jacksoni		E Africa
C. kivuana		Kivu, E Zaire
C. lamottei		Ivory Coast, Togo
C. lanosa		E Zaire, Rwanda
C. lasia		Asia Minor, etc.
C. lasiura		Korea – Ussuri
C. latona		E Zaire
C. lea		Sulawesi
C. leucodon		C, S Europe – Israel
C. levicula		Sulawesi
C. littoralis		Uganda – Cameroun
C. longipes		W Nigeria; swamps; ref. 3.12
C. lucina		Ethiopia; montane
C. luluae		S Zaire
C. luna	Grey-brown musk shrew	Angola – Ethiopia; savanna
C. lusitania		W Sahara
C. macarthuri		Kenya
C. macowi		Mt Nyiro, N Kenya
C. manengubae		Cameroun; ref. 3.7
C. maquassiensis	Maquassie musk shrew	Transvaal, Zimbabwe; *
C. mariquensis	Swamp musk shrew	S Africa – Zambia
C. maurisca		Uganda, Kenya, Tanzania
C. maxi		Java, Flores, etc.; (in C. neglecta?)

C. mindorus		Mindoro, Negros, Philippines
C. miya	Ceylon long-tailed shrew	Sri Lanka
C. monax		E Africa; montane
C. monticola (*bartelsii*)		Malaya, Borneo, Java
C. mutesae		Uganda
C. nana		Somalia, Ethiopia; (in *C. religiosa?*)
C. nanilla		W, E Africa
C. neglecta		Sumatra
C. negrina		Negros I, Philippines
C. nicobarica		Gt Nicobar I, Indian Ocean
C. nigricans		Angola
C. nigripes		Sulawesi
C. nigrofusca		Zaire
C. nimbae		Ivory Coast – Guinea
C. niobe		Ruwenzori, Uganda, SW Ethiopia
C. odorata (*goliath*)		Guinea – Gabon
C. palawanensis		Palawan, Philippines
C. paradoxura		Sumatra; a doubtful species
C. parvacauda		Mindanao, Philippines
C. pasha		Sudan – Mali; savanna
C. pergrisea (? *armenica*)		Asia Minor – Tien Shan; montane
C. phaeura		S Ethiopia
C. picea		Cameroun
C. pitmani		Zambia
C. planiceps		N Uganda, S Sudan
C. poensis		W, C Africa
C. religiosa	Egyptian pygmy shrew	Egypt
C. rhoditis		Sulawesi
C. roosevelti		Uganda, Tanzania – Angola
C. russula (*heljanensis*)		C, S Europe, N Africa
C. sansibarica		Zanzibar, Pemba I
C. serezkyensis		Pamir Mts, etc.; ref. 3.15
C. sibirica		C Asia
C. sicula		Sicily

C. smithi		Ethiopia, Somalia, Senegal
C. somalica		Sudan, Somalia, Ethiopia, Oman
C. suaveolens		SW Europe, N Africa – Korea, China, Taiwan
C. susiana		SW Iran; steppe
C. thalia		Ethiopia; montane; 3.16
C. theresae		Guinea – Ghana
C. turba		Zambia – N Angola
C. usambarae		Usambara Mts, Tanzania; ref. 3.16
C. viaria		Morocco – Senegal –
(*suahelae*)		Kenya; ref. 3.8
(*bolivari*)		
(*tephra*)		
C. vulcani		Mt Cameroun
C. whitakeri		NW Africa
C. wimmeri		Liberia – Gabon
C. xantippe		Kenya, Tanzania
C. yankariensis		N Nigeria, Sudan; ref. 3.10
C. zaodon		Zambia – S Sudan, ? W Africa
C. zaphiri		Kenya, S Ethiopia
C. zarudnyi		Afghanistan, Baluchistan
C. zimmeri		S Zaire

Suncus; (*Pachyura, Podihik*); Africa, S Eurasia; forest, scrub, savanna.

S. ater	Black shrew	Mt Kinabalu, Borneo
S. dayi		S India
S. etruscus	Pygmy white-toothed shrew	Mediterranean – India, Sri Lanka; scrub, etc.
S. hosei		Sarawak, Borneo
S. infinitesimus	Least dwarf shrew	S Africa – Kenya, ? W Africa
S. lixus	Greater dwarf shrew	Kenya – Angola, Transvaal
S. luzoniensis		Luzon, Philippines
S. malayanus		Malaya, Borneo; (in *S. etruscus?*)
S. mertensi		Flores
S. murinus	House shrew	S Asia, (E Africa); commensal
S. occultidens		S Philippines
S. palawanensis		Palawan, Philippines
S. remyi		Gabon
S. stoliczkanus		India, etc.

| *S. varilla* | Lesser dwarf shrew | S Africa – Tanzania, Zaire |

Feroculus
| *F. feroculus* | Kelaart's long-clawed shrew | Sri Lanka; montane forest |

Solisorex
| *S. pearsoni* | Pearson's long-clawed shrew | Sri Lanka; montane |

Paracrocidura
| *P. schoutedeni* | | Cameroun – Ruwenzori |

Sylvisorex; (*Suncus*); Africa, forest, grassland.
S. granti		Cameroun – E Africa; montane
S. howelli		Uluguru Mts, Tanzania; ref. 3.13
S. johnstoni		Cameroun, Gabon, Zaire, Bioco
S. lunaris		C Africa; montane
S. megalura	Climbing shrew	Guinea – Ethiopia – Zimbabwe
S. morio		Mt Cameroun, Bioco
S. ollula		Cameroun – Zaire
S. suncoides		C Africa; montane; (in *S. ollula*?)

Myosorex; (*Surdisorex*); mouse-shrews; C, S Africa; forest.
M. baboulti		E Zaire
M. blarina		C Africa; montane
M. cafer	Dark-footed forest shrew	S Africa, Zimbabwe
M. eisentrauti		Cameroun, Bioco
M. geata		Tanzania
M. longicaudatus	Long-tailed forest shrew	Knysna, S Africa
M. norae		Aberdare Mts, Kenya
M. polli		Kasai, Zaire
M. polulus		Mt Kenya; montane scrub
M. preussi		Mt Cameroun
M. varius	Forest shrew	S Africa

Diplomesodon
| *D. pulchellum* | Piebald shrew | Russian Turkestan; desert |

Anourosorex
A. *squamipes*   Mole-shrew   S China – N Thailand,
                                                      Taiwan; montane forest

Chimarrogale; (*Crossogale*); oriental water shrews; E Asia; montane streams.
C. *himalayica*   Himalayas, Malaya, China,
   (*platycephala*)   Taiwan, Japan
C. *phaeura*   Sumatra, Borneo
C. *styani*   Sichuan, China

Nectogale
N. *elegans*   Elegant water shrew   Sikkim – Shaanxi;
                                                          montane streams

Scutisorex
S. *somereni*   Armoured shrew   C Africa; forest
   (*congicus*)

Family Talpidae

Moles, shrew-moles, desmans; *c*. 30 species; Eurasia, N America; mainly
subterranean in forest and wet grassland, some aquatic; predators on
invertebrates.

Subfamily Uropsilinae

Chinese shrew-moles; 3 species; S China, etc.; montane; ref. 3.14.

Uropsilus; (*Nasillus, Rhynchonax*).
U. *andersoni*   C Sichuan
U. *gracilis*   Sichuan, Yunnan,
                                                         N Burma

U. *soricipes*   Chinese shrew-mole   S China, N Burma;
                                                         montane

Subfamily Desmaninae

Desmans; 2 species; Europe; aquatic.

Desmana
D. *moschata*   Russian desman   European Russia,
                                                         [C Siberia]; rivers; *

Galemys
G. *pyrenaicus*   Pyrenean desman   Pyrenees, Iberia; rivers; *

Subfamily Talpinae

Moles; *c.* 25 species; Eurasia, N America; subterranean; generic and specific classification of Eurasian forms very provisional.

Talpa; Eurasian moles; Eurasia; subterranean in forest and grassland.

T. altaica	Siberian mole	W, C Siberia
T. caeca	Mediterranean mole	S Europe – Caucasus
T. caucasica	Caucasian mole	Caucasus
T. europaea	European mole	Europe, W Siberia
T. romana	Roman mole	Italy, Balkans
T. streeti	Persian mole	NW Iran

Mogera; (*Talpa*).

M. latouchei		SE China, Hainan
M. robusta (*kobeae*)	Large Japanese mole	Korea – Japan
M. wogura (? *coreana*)	Japanese mole	Japan, ? Korea

Parascaptor; (*Talpa*).

P. leucura		Yunnan, Burma, Assam

Scaptochirus; (*Talpa*).

S. moschatus	Short-faced mole	E China

Euroscaptor; (*Talpa*).

E. longirostris (*micrura*)		S China
E. micrura (*klossi*)	Himalayan mole	E Himalayas – Malaya
E. mizura	Japanese mountain mole	Honshu, Japan

Scaptonyx

S. fusicaudus	Long-tailed mole	S China, N Burma; montane

Neurotrichus

N. gibbsii	American shrew-mole	W coast USA; wet forest

Urotrichus; (*Dymecodon*).

U. pilirostris	Lesser Japanese shrew-mole	Japan; montane forest
U. talpoides	Greater Japanese shrew-mole	Japan

Scapanulus
S. *oweni*   Kansu mole   C China; montane forest

Parascalops
P. *breweri*   Hairy-tailed mole   NE USA, SE Canada

Scapanus; western moles; W USA; wet forest, grassland.
S. *latimanus*   Broad-footed mole   California, etc.
S. *orarius*   Coast mole   W USA
S. *townsendii*   Townsend's mole   W coast USA

Scalopus
S. *aquaticus*   Eastern American mole   C, E USA, NE Mexico
   (*inflatus*)
   (*montanus*)

Condylura
C. *cristata*   Star-nosed mole   NE USA, SE Canada

ORDER SCANDENTIA

One family, variously included in the Insectivora or Primates, or associated with the Macroscelidea, but probably better treated as a separate order.

Family Tupaiidae

Tree shrews; *c.* 16 species; SE Asia; forest; arboreal insectivores.

Tupaia
T. *dorsalis*   Striped tree shrew   Borneo
T. *glis*   Common tree shrew   Nepal – S China – Malaya,
   (*belangeri*)   Sumatra, Java, Borneo;
   (*chinensis*)   possibly more than one
                                                 species
T. *gracilis*   Slender tree shrew   Borneo
T. *javanica*   Javan tree shrew   Java, Sumatra
T. *minor*   Pygmy tree shrew   Malaya, Sumatra, Borneo
T. *montana*   Mountain tree shrew   Borneo; montane
T. *nicobarica*   Nicobar tree shrew   Nicobar Is, Indian Ocean
T. *palawanensis*   Palawan tree shrew   Palawan, etc., Philippines
T. *picta*   Painted tree shrew   Borneo
T. *splendidula*   Ruddy tree shrew   S Borneo; lowland
                        (Red-tailed tree shrew)
T. *tana*   Large tree shrew   Borneo, Sumatra

Anathana
A. ellioti	Madras tree shrew	S, C India

Dendrogale
D. melanura	Bornean smooth-tailed tree shrew	Borneo; montane
D. murina	Northern smooth-tailed tree shrew	S Indochina

Urogale
U. everetti	Philippine tree shrew	Mindanao, Philippines

Ptilocercus
P. lowii	Pen-tailed tree shrew	Malaya, Sumatra, W, N Borneo

ORDER DERMOPTERA

One family.

Family Cynocephalidae

Flying lemurs (colugos); 2 species; SE Asia; forest; arboreal, gliding, herbivores.

Cynocephalus; (*Galeopithecus*).
C. variegatus	Malayan flying lemur	Indochina – Java, Borneo
C. volans	Philippine flying lemur	Philippines

ORDER CHIROPTERA

Bats; *c.* 988 species; worldwide.

SUBORDER MEGACHIROPTERA

Family Pteropodidae

Fruit bats, flying foxes; *c.* 174 species; Old-world tropics and subtropics: Africa, S Asia to Australia, W Pacific Islands; mainly forest, rarely in caves; frugivorous, nectarivorous; ref. 4.9.

Subfamily Pteropodinae
Eidolon

E. helvum	Straw-coloured fruit bat	Africa S of Sahara, Ethiopia, SW Arabia, Madagascar

Rousettus; rousettes; Africa, S Asia – Solomon Is.
Subgenus *Rousettus*

R. aegyptiacus	Egyptian rousette	S Africa – Senegal, Ethiopia, Egypt – Lebanon – Pakistan, Cyprus
R. amplexicaudatus (stresemanni)	Geoffroy's rousette	S Burma – Solomon Is, Philippines; refs. 4.10, 11, 12
R. celebensis	Sulawesi rousette	Sulawesi, Sanghir Is
R. lanosus	Ruwenzori long-haired rousette	S Ethiopia – Tanzania, E Zaire
R. leschenaulti	Leschenault's rousette	Pakistan – Vietnam – S China; Sri Lanka, Java, Bali
R. madagascariensis	Madagascar rousette	Madagascar
R. obliviosus		Grande Comoro I, Anjouan I, Comoro Is
R. spinalatus		N Sumatra, Borneo; refs. 4.11, 13, 14

Subgenus *Lissonycteris*

R. angolensis	Bocage's fruit bat	Guinea – Kenya, Angola, Zimbabwe, Zambia, Mozambique

Myonycteris

M. brachycephala	Sào Tomé collared fruit bat	Saò Tomé I, Gulf of Guinea
M. relicta		SE Kenya, NE Tanzania; ref. 4.15
M. torquata	Little collared fruit bat	Sierra Leone – Angola, Zambia

Boneia

B. bidens		N Sulawesi

Pteropus; flying foxes; Madagascar, SE Asia, N Australia, islands of Indian Ocean and W Pacific.

P. admiralitatum	Admiralty flying fox	Admiralty Is – Solomons

P. alecto	Central flying fox (Black flying fox)	Sulawesi – S New Guinea, NW, N, NE Australia
P. anetianus		Vanuatu, Banks Is
P. argentatus	Silvery flying fox	? Amboina I; ref. 4.16
P. brunneus		Percy I, off E Queensland; perhaps applied to a vagrant P. hypomelanus
P. caniceps	Ashy-headed flying fox	Sulawesi, N Moluccas, etc.
P. chrysoproctus	Amboina flying fox	Sanghir Is; S Moluccas
P. conspicillatus	Spectacled flying fox	N Moluccas, New Guinea – NE Queensland
P. dasymallus	Ryukyu flying fox	S Japan, Ryukyu Is, Taiwan
P. faunulus		Car Nicobar I, Indian Ocean
P. fundatus		Banks Is, Vanuatu
P. giganteus	Indian flying fox	India – Burma, W China, Sri Lanka, Maldive Is, Andaman Is
P. gilliardi	Gilliard's flying fox	New Britain, Bismarck Arch.
P. griseus	Grey flying fox	Timor – Sulawesi, ? Luzon
P. howensis		Ontong Java Atoll, Solomons
P. hypomelanus (? santacrucis)	Small flying fox	S Burma – Solomons, Philippines, C Maldive Is
P. insularis		Truk Is, Caroline Is
P. intermedius		S Burma
P. leucopterus		Luzon, Philippines
P. livingstonii	Comoro black flying fox	Anjouan I, Moheli I, Comoro Is; ref. 4.17
P. lombocensis	Lombok flying fox	Lesser Sunda Is
P. lylei	Lyle's flying fox	Thailand, Indochina
P. macrotis	Big-eared flying fox	S New Guinea, Aru Is
P. mahaganus	Lesser flying fox	Ysabel I, Bougainville I, Solomons
P. mariannus (loochooensis) (pelewensis) (ualanus) (yapensis)	Marianas flying fox	Mariana Is, Palau Is, Caroline Is, Ryukyu Is; *
P. mearnsi		Mindanao, Basilan, S Philippines

P. melanopogon	Black-bearded flying fox	Sanghir Is; S Moluccas
P. melanotus (*satyrus*)		Andaman Is, Nicobar Is – Christmas I, Indian Ocean
P. molossinus		E Caroline Is
P. neohibernicus (*sepikensis*)	Bismarck flying fox	New Guinea, Bismarck Arch.
P. niger	Greater Mascarene flying fox	† Reunion, Mauritius; ref. 4.17; *
P. nitendiensis		Santa Cruz Is, SW Pacific
P. ocularis	Ceram flying fox	Buru, Ceram Is, S Moluccas
P. ornatus		New Caledonia; Loyalty Is
P. personatus	Masked flying fox	Sulawesi, N Moluccas
P. phaeocephalus		Mortlock I (= Tauu I), C Caroline Is
P. pilosus		Palau Is
P. pohlei	Geelvink Bay flying fox	Japen I, W New Guinea
P. poliocephalus	Grey-headed flying fox	E Australia, vagrant Tasmania
P. pselaphon		Bonin Is; Volcano Is, (S of Japan)
P. pumilus (*balutus*) (*tablasi*)		Philippines; ref. 4.18
P. rayneri (*cognatus*)	Solomon flying fox	Solomon Is
P. rodricensis	Rodriguez flying fox	Rodriguez I; † Round I; ref. 4.17; *
P. rufus	Madagascar flying fox	Madagascar
P. samoensis	Samoa flying fox	Fiji; Samoa
P. scapulatus	Little red flying fox	W, N, E Australia; S New Guinea; vagrant New Zealand
P. seychellensis	Seychelles flying fox	Comoro Is, Aldabra I, Seychelles; Mafia I
P. speciosus		Philippines, etc.
P. subniger†		Reunion, Mauritius; extinct; ref. 4.17
P. temmincki	Temminck's flying fox	S Moluccas, Bismarck Arch.
P. tokudae	Guam flying fox	Guam I; *
P. tonganus	Insular flying fox	Karkar I (NE New Guinea) – Samoa, Cook Is

P. tuberculatus		Vanikoro I, Santa Cruz Is
P. vampyrus	Large flying fox	S Burma – Java, Philippines, Borneo, Timor
? *P. vanikorensis*		Vanikoro Is, Santa Cruz Is
P. vetulus		New Caledonia
P. voeltzkowi	Pemba flying fox	Pemba I, off Tanzania
P. woodfordi	Least flying fox	Solomons

Pteralopex

P. acrodonta		Taveuni I, Fiji Is
P. anceps		Bougainville, Choiseul Is, Solomons
P. atrata	Cusp-toothed flying fox	Ysabel, Guadalcanal Is, Solomons

Acerodon; Philippines – Lesser Sundas.

A. celebensis (*argentatus*) (*arquatus*)	Sulawesi flying fox	Sulawesi; ref. 4.16
A. humilis	Talaud flying fox	Talaud Is, N Moluccas
A. jubatus		Philippines
A. leucotis		Busuanga I, Palawan, Philippines; ref. 4.16
A. lucifer		Panay I, C Philippines
A. mackloti	Sunda flying fox	Lesser Sunda Is

Neopteryx

N. frosti	Small-toothed fruit bat	W Sulawesi

Styloctenium

S. wallacei	Stripe-faced fruit bat	Sulawesi

Dobsonia; naked-backed fruit bats; Philippines – Australia.

D. beauforti		Biak I, Owii I, Waigeo I, off NW New Guinea
D. emersa		Biak I, Owii I, off NW New Guinea; ref. 4.19
D. exoleta	Sulawesi naked-backed bat	Sulawesi
D. inermis	Solomons naked-backed bat	Solomons
D. minor	Lesser naked-backed bat	C, W New Guinea

D. moluccense (? *anderseni*)	Greater naked-backed bat	Moluccas, New Guinea, S Bismarck Archipelago, Trobriand, D'Entrecasteaux, Louisiade Is, N Queensland; refs. 4.12, 19
D. peronii	Western naked-backed bat	Lesser Sunda Is, Timor
D. praedatrix	New Britain naked-backed bat	Bismarck Archipelago
D. remota	Trobriand naked-backed bat	Trobriand Is
D. viridis (? *crenulata*)	Greenish naked-backed bat	Negros I, Philippines; Sulawesi; Moluccas; ref. 4.12

Plerotes

P. anchietae	Anchieta's fruit bat	Angola, S Zaire, Zambia

Hypsignathus

H. monstrosus	Hammer-headed fruit bat	Gambia – Ethiopia – Zaire, NE Angola

Epomops; epauletted fruit bats.

E. buettikoferi	Büttikofer's fruit bat	Guinea – Ghana; ? Nigeria
E. dobsoni	Dobson's fruit bat	W, C Angola – S Zaire, Rwanda, NE Botswana, Zambia, Tanzania
E. franqueti	Franquet's fruit bat (Singing fruit bat)	Ivory Coast – S Sudan – NW Tanzania, Angola

Epomophorus; epauletted fruit bats; Africa S of Sahara.

E. angolensis	Angolan epauletted fruit bat	SW Angola, NW Namibia
E. anurus	Eastern epauletted fruit bat	Senegal – Ethiopia – Tanzania; (in *E. labiatus* ?)
E. crypturus	Peters' epauletted fruit bat	Angola – S Tanzania – S Africa
E. gambianus	Gambian epauletted fruit bat	Senegal – S Ethiopia
E. labiatus	Little epauletted fruit bat	N Ethiopia, S Sudan, ? Congo Rep.

E. minor		S Ethiopia – Zambia, Malawi, Zanzibar
E. pousarguesi	Pousargue's epauletted fruit bat	Cent. African Rep.
E. reii	Garua epauletted fruit bat	Cameroun
E. wahlbergi	Wahlberg's epauletted fruit bat	Somalia – S Africa, Angola, Zaire, Cameroun

Micropteropus; dwarf epauletted fruit bats.

M. grandis	Sanborn's epauletted fruit bat	NE Angola; Congo Rep.
M. intermedius	Hayman's epauletted fruit bat	N Angola, SE Zaire
M. pusillus	Dwarf epauletted fruit bat	Senegal – Ethiopia, Angola – Zambia

Nanonycteris

N. veldkampii	Veldkamp's dwarf fruit bat	Guinea – Congo Rep.

Scotonycteris

S. ophiodon	Pohle's fruit bat	Liberia – Congo Rep.
S. zenkeri	Zenker's fruit bat	Liberia – E Zaire

Casinycteris

C. argynnis	Short-palate fruit bat	Cameroun – E, NE Zaire

Cynopterus

C. archipelagus		Polillo I, off Luzon, Philippines
C. brachyotis	Lesser dog-faced fruit bat	S China, Burma – Java, Borneo, Philippines, Sulawesi, Sri Lanka, Andamans, Nicobars
C. horsfieldi	Horsfield's fruit bat	S Thailand – Java, Borneo
C. minor		Sulawesi; ref. 4.12
C. sphinx	Short-nosed fruit bat	India – S China – Java, ? Borneo
C. titthaecheilus		Sumatra, Krakatoa I, Nias I, Java; ? Timor; ref. 4.12

Megaerops

M. ecaudatus	Tail-less fruit bat	Thailand – Sumatra, Borneo

M. kusnotoi   Java

M. niphanae   Thailand, ? NE India, ? Vietnam; refs. 4.12, 20

M. wetmorei   Mindanao, Philippines

Ptenochirus

P. jagori   Philippines

P. minor   Mindanao, Palawan

Dyacopterus

D. brooksi   Sumatra

D. spadiceus   Dyak fruit bat   Malaya – Borneo, Luzon

Chironax

C. melanocephalus   Black-capped fruit bat   S Thailand – Java, Borneo, Sulawesi; ref. 4.14

Latidens

L. salimalii   S India

Penthetor

P. lucasi   Lucas' short-nosed fruit bat   Malaya, Borneo

Thoopterus

T. nigrescens   Swift fruit bat   N Sulawesi, Morotai I, N Moluccas; ? Luzon

Aproteles

A. bulmerae   E New Guinea

Sphaerias

S. blanfordi   Blanford's fruit bat   NE India, S Tibet – NW Thailand, SW China

Balionycteris

B. maculata   Spotted-winged fruit bat   S Thailand, Malaya, Borneo

Aethalops

A. alecto   Pygmy fruit bat   Malaya, Sumatra, Java, Borneo; ref. 4.12

Haplonycteris

H. fischeri   Mindoro, Luzon, Philippines

Alionycteris
A. *paucidentata*   Mindanao, Philippines

Otopteropus
O. *cartilagonodus*   Luzon, Philippines

Subfamily Harpyionycterinae

Harpyionycteris
H. *celebensis*   Sulawesi
H. *whiteheadi*   Harpy fruit bat   Philippines

Subfamily Nyctimeninae

Nyctimene; tube-nosed fruit bats; New Guinea, etc.
N. *aello*   Broad-striped tube-nosed   New Guinea
                          bat
N. *albiventer*   Common tube-nosed bat   N Moluccas, New Guinea,
                                                       Admiralty Is – Solomon
                                                       Is, NE Australia
N. *celaeno*   W, NW New Guinea;
                                                       ref. 4.12
N. *cephalotes*   Pallas' tube-nosed bat   Sulawesi, Timor –
                                                       NW New Guinea,
                                                       Admiralty Is
N. *cyclotis*   Round-eared tube-nosed   New Guinea;
                          bat   E New Britain
N. *draconilla*   Papua New Guinea;
                                                       ref. 4.12
N. *major*   Greater tube-nosed bat   Bismarck Arch.,
                                                       Trobriand, Louisiade,
                                                       D'Entrecasteaux,
                                                       Solomon Is
N. *malaitensis*   Malaita tube-nosed bat   Malaita I, E Solomon Is
N. *masalai*   New Ireland; ref. 4.21
N. *minutus*   Lesser tube-nosed bat   Sulawesi, Buru
N. *rabori*   Negros I, Philippines;
                                                       ref. 4.22
N. *robinsoni*   Queensland tube-nosed bat   NE Australia
N. *sanctacrucis*   Santa Cruz Is
N. *vizcaccia*   New Guinea; ref. 4.21

Paranyctimene

P. raptor   Lesser tube-nosed bat   New Guinea

Subfamily Macroglossinae

Eonycteris; dawn fruit bats; SE Asia.

E. major   Borneo; Mentawei Is
E. robusta   Philippines; (in *E. major*?)
E. rosenbergi   Sulawesi dawn bat   N Sulawesi; ref. 4.23
E. spelaea   Dawn bat   Burma – Java, Sumba,
                   (Cave fruit bat)   Sulawesi, Philippines,
                                                      Timor; refs. 4.12, 24

Megaloglossus

M. woermanni   African long-tongued fruit   Guinea – Uganda –
                   bat   N Angola

Macroglossus; long-tongued fruit bats; SE Asia; ref. 4.12.

M. fructivorus   Mindanao, Philippines; (in
                                                     M. minimus?)

M. minimus   Common long-tongued   Thailand, Indochina,
   (*lagochilus*) fruit bat   Philippines – Solomons,
                                                      NW, N Australia

M. sobrinus   Hill long-tongued fruit bat   NE India Java, Bali

Syconycteris

S. australis   Common blossom bat   S Moluccas, New Guinea,
   (*naias*)   (Southern blossom bat)   Bismarck Arch.,
   (*crassa*)   Trobriand I,
                                                      D'Entrecasteaux Is,
                                                      NE Australia; ref. 4.12

S. carolinae   Halmahera, N Moluccas;
                                                     ref. 4.23

S. hobbit   New Guinea; ref. 4.25

Melonycteris; (*Nesonycteris*).
Subgenus *Melonycteris*

M. melanops   Black-bellied fruit bat   Bismarck Arch.,
                                                     New Guinea?

Subgenus *Nesonycteris*
M. aurantius   Orange fruit bat   Florida, Choiseul Is,
                                                     Solomons

M. woodfordi	Woodford's fruit bat	Solomons

Notopteris

N. macdonaldii	Long-tailed fruit bat	Vanuatu, New Caledonia, Fiji; ? Caroline Is

SUBORDER MICROCHIROPTERA

Family Rhinopomatidae

Mouse-tailed bats (rat-tailed bats, long-tailed bats); 3 species; Morocco, Senegal – Thailand, Sumatra; mainly desert and steppe; insectivorous.

Rhinopoma

R. hardwickii	Lesser mouse-tailed bat	Morocco, Mauretania, Nigeria – Kenya – Thailand
R. microphyllum	Greater mouse-tailed bat	Senegal – India; Sumatra
R. muscatellum		S Arabia – W Pakistan

Family Emballonuridae

Sheath-tailed bats (sac-winged bats, pouched bats, ghost bats); *c.* 50 species; tropics and subtropics of world; insectivorous.

Subfamily Emballonurinae

Emballonura; Old-world sheath-tailed bats; Madagascar, S Burma – Pacific Islands.

E. alecto (*rivalis*)	Philippine sheath-tailed bat	Philippines, Borneo – S Moluccas
E. atrata	Peters' sheath-tailed bat	Madagascar
E. beccarii	Beccari's sheath-tailed bat	New Guinea, etc.
E. dianae	Rennell Island sheath-tailed bat	New Guinea, New Ireland, Malaita, Rennell Is, Solomons; ref. 4.26
E. furax	Greater sheath-tailed bat	New Guinea, New Ireland; ref. 4.26
E. monticola	Lesser sheath-tailed bat	S Burma – Sulawesi
E. nigrescens		Sulawesi – Solomons
E. raffrayana	Raffray's sheath-tailed bat	Ceram, NW New Guinea, Solomons, Santa Cruz Is; ref. 4.26

E. semicaudata
  (*rotensis*)
		Vanuatu – Mariana,
		  Palau Is, Samoa,
		  Fiji

E. sulcata		Caroline Is
Coleura		
C. afra	African sheath-tailed bat	Africa, Aden
C. seychellensis	Seychelles sheath-tailed bat	Seychelles; ? Zanzibar
Rhynchonycteris		

R. naso	Long-nosed bat
	  (Tufted bat)
		S Mexico – Bolivia, Brazil;
		  Trinidad

Saccopteryx		

S. bilineata	Greater white-lined bat
	  (Sac-winged bat)
		W, E Mexico – Bolivia,
		  Brazil; Trinidad

S. canescens		Colombia – Peru, Brazil
S. gymnura		Brazil; ? Venezuela
S. leptura	Lesser white-lined bat	W Mexico – Peru, Brazil; Trinidad
Cormura		
C. brevirostris	Wagner's sac-winged bat	Nicaragua – Peru, Brazil
Peropteryx		

P. kappleri	Greater sac-winged bat
	  (Greater dog-like bat)
		S Mexico – E Peru,
		  Surinam, S, E Brazil

P. macrotis
  (*trinitatis*)
	Lesser sac-winged bat
	  (Lesser dog-like bat)
		S Mexico – Peru, Brazil;
		  Tobago, Grenada,
		  Trinidad

Peronymus		
P. leucopterus		Venezuela, Guyana, Surinam, E Peru, Brazil
Centronycteris		

C. maximiliani	Shaggy-haired bat
	  (Thomas's bat)
		S Mexico – Ecuador,
		  ? E Peru, Brazil

Balantiopteryx; least sac-winged bats.		
B. infusca		Ecuador
B. io	Thomas's least sac-winged bat	S Mexico – Guatemala
B. plicata	Peters' bat	N Mexico – Costa Rica

Taphozous; pouched bats, tomb bats, sheath-tailed bats; Africa, S Asia, Australia.
Subgenus *Taphozous*

T. australis	Gould's pouched bat	New Guinea, NE Queensland
T. georgianus	Sharp-nosed pouched bat	W, N Australia
T. hildegardeae	Hildegarde's tomb bat	Kenya, NE Tanzania, Zanzibar
T. hilli	Hill's pouched bat	W, N Australia; ref. 4.27
T. kapalgensis	White-striped sheath-tailed bat	N Australia; ref. 4.28
T. longimanus	Long-winged tomb bat	India – Java, Flores
T. mauritianus	Mauritian tomb bat	Africa S of Sahara, Madagascar, Aldabra, Mauritius, Reunion, Assumption
T. melanopogon	Black-bearded tomb bat	India – Java, Lesser Sundas, Borneo, Philippines
T. perforatus	Egyptian tomb bat	Senegal – Somalia, Mozambique, Zimbabwe – India
T. solifer		? China
T. theobaldi	Theobald's tomb bat	India – Vietnam; Java

Subgenus *Saccolaimus*

T. flaviventris	Yellow-bellied pouched bat	Australia
T. mixtus	Troughton's pouched bat	S, E New Guinea, N Queensland
T. peli	Pel's pouched bat	Liberia – NE Angola, Kenya
T. pluto (*capito*)		Philippines; (in *T. saccolaimus*?)
T. saccolaimus (*nudicluniatus*)		India – Java, Borneo, Timor, New Guinea, Solomon Is, N, NE Australia

Subgenus *Liponycteris*

T. hamiltoni	Hamilton's tomb bat	Sudan, NW Kenya, ? S Chad; ref. 4.29
T. nudiventris	Naked-rumped tomb bat	Senegal – Somalia, Tanzania, Israel, Arabia, E Afghanistan – Burma

Subfamily Diclidurinae

Diclidurus; ghost bats (white bats).

D. albus		E Peru, Venezuela, Surinam, Brazil, Trinidad
D. ingens		Colombia – Guyana, NW Brazil
D. scutatus		Venezuela – Surinam, Peru, Brazil
D. virgo	White bat (Northern ghost bat)	W Mexico – Colombia, Venezuela

Depanycteris; (*Drepanycteris*); possibly a subgenus of *Diclidurus*.

D. isabella		Venezuela, Brazil

Cyttarops

C. alecto	Short-cared bat	Costa Rica, Nicaragua, Guyana, Brazil

Family Craseonycteridae

One species.

Craseonycteris

C. thonglongyai	Hog nosed bat (Butterfly bat)	S Thailand

Family Nycteridae

Slit-faced bats, hollow-faced bats, hispid bats; *c*. 14 species; Africa, SW, SE Asia to Java; chiefly forest and drier areas with trees; insectivorous, (carnivorous).

Nycteris

N. arge	Bate's slit-faced bat	Sierra Leone – SW Sudan, W Kenya, NW Angola, Bioco
N. gambiensis	Gambian slit-faced bat	Senegal – Ghana, Togo
N. grandis	Large slit-faced bat	Senegal – Kenya, Tanzania, Mozambique, Zimbabwe; ref. 4.29
N. hispida	Hairy slit-faced bat	Senegal – Ethiopia – S Africa
N. javanica	Javan slit-faced bat	Java, Bali, Timor

N. macrotis	Dobson's slit-faced bat (Greater slit-faced bat)	Gambia – Somalia – Zimbabwe, Mozambique
N. madagascariensis		Madagascar
N. major	Ja slit-faced bat	Benin – E, S Zaire, Gabon
N. nana	Dwarf slit-faced bat	Ghana – W Kenya – NE Angola
N. parisii	Parisi's slit-faced bat	Cameroun, Ethiopia, Somalia; ? S Tanzania
N. thebaica	Egyptian slit-faced bat	Africa S of Sahara, Morocco, Egypt, Israel, Arabia
N. tragata		S Burma – Sumatra, Borneo
N. vinsoni	Vinson's slit-faced bat	Mozambique
N. woodi	Wood's slit-faced bat	Tanzania, Zambia, Zimbabwe

Family Megadermatidae

False vampire bats, yellow-winged bats; *c.* 5 species; Africa, SE Asia, Australia; forest, savanna; insectivores, carnivores.

Megaderma; (*Lyroderma*).

M. lyra	Greater false vampire	E Afghanistan – S China, Malaya; Sri Lanka
M. spasma	Lesser false vampire	India – Java, Sulawesi, Philippines – N Moluccas, Sri Lanka

Macroderma

M. gigas	Australian false vampire (Ghost bat)	W, C, N Australia; *

Cardioderma

C. cor	Heart-nosed bat	Ethiopia – N Tanzania, Zanzibar

Lavia

L. frons	Yellow-winged bat	Senegal – Somalia – Zambia

Family Rhinolophidae

Horseshoe bats; *c.* 68 species; Old-world tropics and temperate regions east to Australia; insectivores.

Rhinolophus; (*Rhinomegalophus*).

R. acuminatus	Acuminate horseshoe bat	Thailand – Java – Lombok, Borneo, Palawan
R. adami		Congo Republic
R. affinis	Intermediate horseshoe bat	N India – S China – Java, Lesser Sundas
R. alcyone	Halcyon horseshoe bat	Senegal – NE Zaire, Gabon
R. anderseni		Palawan, ? Luzon, Philippines
R. arcuatus	Arcuate horseshoe bat	Philippines, Borneo, Sumatra, Moluccas, New Guinea
R. blasii	Blasius' horseshoe bat	N Africa, Italy – Afghanistan, Ethiopia – Transvaal, Mozambique
R. bocharicus		NE Iran, N Afghanistan
R. borneensis (*importunus*)	Bornean horseshoe bat	Kampuchea, Borneo, Java; ref. 4.12
R. capensis	Cape horseshoe bat	Zambia, Zimbabwe, S Africa
R. celebensis (*javanicus*) (*madurensis*) (*parvus*)	Sulawesi horseshoe bat	S Sulawesi, Java, Timor; ref. 4.12
R. clivosus	Geoffroy's horseshoe bat	Arabia – Algeria, Ethiopia – Zambia, Mozambique
R. coelophyllus	Croslet horseshoe bat (Peters' horseshoe bat)	Burma – Malaya
R. cognatus		Andaman Is
R. cornutus	Little Japanese horseshoe bat	Japan, Ryukyu Is; ? E China
R. creaghi	Creagh's horseshoe bat	Borneo, Java, Timor
R. darlingi	Darling's horseshoe bat	Tanzania – Namibia, Transvaal, Mozambique
R. deckenii	Decken's horseshoe bat	Kenya, Tanzania
R. denti	Dent's horseshoe bat	Guinea; Mozambique – Namibia, S Africa
R. eloquens		S Sudan – S Somalia – N Tanzania
R. euryale	Mediterranean horseshoe bat	Portugal, Morocco – Iran

R. euryotis	Broad-eared horseshoe bat	Sulawesi, Moluccas, New Guinea, New Britain, Aru Is
R. feae		Burma
R. ferrumequinum	Greater horseshoe bat	Britain, Morocco – N India, Japan
R. fumigatus	Rüppell's horseshoe bat	Senegal – N Ethiopia – S Africa
R. gracilis		SE India; status doubtful
R. guineensis		Senegal – Sierra Leone; ref. 4.30
R. hildebrandtii	Hildebrandt's horseshoe bat	Ethiopia, Somalia – Botswana, Transvaal
R. hipposideros	Lesser horseshoe bat	Britain, Ireland – N India, NE Africa
R. imaizumii		Iriomote I, Ryukyu Is; ref. 4.31
R. inops		Mindanao, Philippines
R. keyensis	Insular horseshoe bat	Moluccas, Wetter I, Kei Is
R. landeri	Lander's horseshoe bat	Senegal – Somalia – Angola, Transvaal, Mozambique
R. lepidus (*refulgens*)	Blyth's horseshoe bat	Afghanistan – S China – Sumatra; ref. 4.31
R. luctus	Woolly horseshoe bat	N India – S China – Java, Borneo
R. maclaudi (*hilli*) (*ruwenzorii*)	Maclaud's horseshoe bat	Guinea, E Zaire, W Uganda, Rwanda
R. macrotis (*hirsutus*)	Big-eared horseshoe bat	Nepal – Malaya, Sumatra, Philippines
R. malayanus	N Malayan horseshoe bat	Thailand – Vietnam, Malaya
R. marshalli	Marshall's horseshoe bat	Thailand
R. megaphyllus	Southern horseshoe bat (Eastern horseshoe bat)	SE New Guinea, New Ireland, New Britain, E Australia
R. mehelyi	Mehely's horseshoe bat	Portugal, Morocco – Afghanistan
R. mitratus		C India; status doubtful
R. monoceros		Taiwan
R. nereis		Anamba Is, Natuna Is
R. osgoodi		Yunnan, China
R. paradoxolophus	Bourret's horseshoe bat	Thailand, Vietnam

R. pearsoni	Pearson's horseshoe bat	NE India – S China, Indochina
R. petersi		India; (in *P. rouxii*?)
R. philippinensis	Philippines horseshoe bat (Large-eared horseshoe bat)	Philippines, Borneo – Sulawesi, Timor, NE Queensland
R. pusillus (*blythi*) (*minutillus*)	Least horseshoe bat	NE India, S China – Java, Borneo
R. rex		S China
R. robinsoni	Peninsular horseshoe bat	S Thailand, Malaya
R. rouxii		India, S China, Vietnam, Sri Lanka
R. rufus		Philippines
R. sedulus	Lesser woolly horseshoe bat	Malaya, Borneo
R. shameli	Shamel's horseshoe bat	Burma – Indochina, Malaya
R. silvestris		Gabon, Congo Rep.
R. simplex	Lombok horseshoe bat	Lesser Sunda Is
R. simulator	Bushveld horseshoe bat	Guinea – Ethiopia – Transvaal, etc.
R. stheno	Lesser brown horseshoe bat	Thailand – Java
R. subbadius		N India – Vietnam
R. subrufus		Philippines
R. swinnyi	Swinny's horseshoe bat	S Zaire, Tanzania – S Africa
R. thomasi	Thomas's horseshoe bat	Burma, Yunnan – Indochina
R. toxopei	Buru horseshoe bat	Buru I, S Moluccas
R. trifoliatus	Trefoil horseshoe bat	N India – Java, Borneo
R. virgo		S Philippines
R. yunanensis	Dobson's horseshoe bat	NE India, Yunnan, Thailand

Family Hipposideridae

Old-world leaf-nosed bats, trident bats; *c*. 63 species; tropics and subtropics of Africa, S Asia – Philippines, Vanuatu, N Australia; forest, savanna; insectivores.

Hipposideros; (*Syndesmotis*)

H. abae	Aba leaf-nosed bat	Guinea – S Sudan, Uganda
H. armiger	Himalayan leaf-nosed bat	N India – Malaya, Taiwan
H. ater	Dusky leaf-nosed bat	India – Java – NW, N Australia; Philippines

H. beatus	Dwarf leaf-nosed bat	Liberia – SW Sudan, Gabon
H. bicolor	Bicolored leaf-nosed bat	India – Java, Sulawesi, Timor, Philippines
H. breviceps		N Pagi I, Mentawei Is
H. caffer	Sundevall's leaf-nosed bat	Africa, Arabia
H. calcaratus (*cupidus*)	Spurred leaf-nosed bat	New Guinea, Bismarck Arch. – Solomon Is; ref. 4.32
H. camerunensis	Greater cyclops bat	Cameroun, E Zaire
H. cervinus	Gould's leaf-nosed bat	Malaya, Sumatra, Philippines – Vanuatu, NE Australia; ref. 4.33
H. cineraceus	Least leaf-nosed bat	Pakistan – Malaya, Borneo
H. commersoni	Commerson's leaf-nosed bat	Gambia – Somalia – Angola, Mozambique; Madagascar
H. coronatus		Mindanao, Philippines
H. coxi	Cox's leaf-nosed bat	Sarawak, Borneo
H. crumeniferus	Timor leaf-nosed bat	Timor I; status doubtful
H. curtus	Short-tailed leaf-nosed bat	Cameroun, ? Nigeria, Bioco
H. cyclops	Cyclops leaf-nosed bat	Guinea – Gabon – SW Kenya; Bioco
H. diadema	Diadem leaf-nosed bat	Burma – Java, Philippines, Borneo, New Guinea, Solomons, N Australia
H. dinops	Fierce leaf-nosed bat	Sulawesi, Peling I, Solomons
H. doriae		Sarawak, Borneo
H. durgadasi		C India; refs. 4.34, 35
H. dyacorum	Dyak leaf-nosed bat	Borneo
H. fuliginosus	Sooty leaf-nosed bat	Guinea – Cameroun – Zaire
H. fulvus	Fulvous leaf-nosed bat	Afghanistan – India – Sri Lanka
H. galeritus	Cantor's leaf-nosed bat (Fawn leaf-nosed bat)	Sri Lanka, India – Java, Borneo; ref. 4.33
H. halophyllus		Thailand; ref. 4.36
H. inexpectatus	Crested leaf-nosed bat	N Sulawesi
H. jonesi	Jones' leaf-nosed bat	Guinea – Nigeria
H. lamottei		Mt Nimba, Guinea; ref. 4.37

H. lankadiva		India, Sri Lanka
H. larvatus	Horsfield's leaf-nosed bat	Bangladesh, S China – Java, Sumba, Borneo
H. lekaguli	Lekagul's leaf-nosed bat	S Thailand, Malaya
H. lylei	Shield-faced leaf-nosed bat	Burma – Malaya
H. maggietaylorae	Maggie's leaf-nosed bat	New Guinea, Bismarck Arch.; ref. 4.32
H. marisae	Aellen's leaf-nosed bat	Ivory Coast – Guinea
H. megalotis	Big-eared leaf-nosed bat	Ethiopia, Kenya
H. muscinus	Fly River leaf-nosed bat	Papua New Guinea
H. nequam	Malay leaf-nosed bat	Malaya
H. obscurus		Philippines
H. papua	Geelvink Bay leaf-nosed bat	Biak I, W New Guinea
H. pratti	Pratt's leaf-nosed bat	SW China, Vietnam
H. pygmaeus		Philippines
H. ridleyi	Ridley's leaf-nosed bat	Malaya, Borneo; ref. 4.12; *
H. ruber	Noack's leaf-nosed bat	Senegal – Ethiopia – Angola
H. sabanus	Sabah leaf-nosed bat	Malaya, Sumatra, Borneo
H. schistaceus		India
H. semoni	Semon's leaf-nosed bat	E New Guinea, NE Queensland
H. speoris	Schneider's leaf-nosed bat	India, Sri Lanka
H. stenotis	Narrow-eared leaf-nosed bat	NW, N Australia
H. turpis	Lesser leaf-nosed bat	S Thailand, Ryukyu Is
H. wollastoni	Wollaston's leaf-nosed bat	S New Guinea
Asellia		
A. patrizii	Patrizi's trident bat	Ethiopia
A. tridens	Trident bat	Morocco, Senegal – Pakistan
Aselliscus		
A. stoliczkanus	Stoliczka's trident bat	Burma, S China – Indochina, Penang I
A. tricuspidatus	Dobson's trident bat	Moluccas – New Hebrides
Anthops		
A. ornatus	Flower-faced bat	Solomon Is

Cloeotis; ref. 4.38.

C. percivali	Short-eared trident bat (Percival's trident bat)	Kenya – SE Botswana – Mozambique, Transvaal

Rhinonycteris; (*Rhinonicteris*); ref. 4.38.

R. aurantius	Orange leaf-nosed bat	NW, N Australia

Triaenops; ref. 4.38.

T. furculus	Trouessart's trident bat	Madagascar, Aldabra, Picard, Cosmoledo Is
T. persicus (? *humbloti*) (? *rufus*)	Persian trident bat	Congo Rep., Mozambique – S Arabia – Pakistan; Madagascar

Coelops

C. frithii	Tail-less leaf-nosed bat	India, S China – Java, Bali, Taiwan
C. robinsoni (*hirsutus*)	Malayan tail-less leaf-nosed bat	S Thailand, Malaya, Borneo, Mindoro, Philippines

Paracoelops

P. megalotis		Vietnam

Family Noctilionidae

Bulldog bats (hare-lipped bats, mastiff bats, fisherman bats); 2 species; New-world tropics; insectivorous, piscivorous.

Noctilio

N. albiventris	Lesser bulldog bat	Honduras – N Argentina
N. leporinus	Greater bulldog bat (Fisherman bat)	W, S Mexico – N Argentina, Antilles, Trinidad, S Bahamas

Family Mormoopidae

Naked-backed bats, moustached bats, ghost-faced bats; *c.* 8 species; S USA – Brazil, Greater and Lesser Antilles; insectivorous.

Pteronotus

Subgenus *Pteronotus*

P. davyi	Davy's naked-backed bat	Mexico – Peru, Brazil, Lesser Antilles, Trinidad
P. gymnonotus (*suapurensis*)	Big naked-backed bat	S Mexico – E Peru, Brazil

Subgenus *Chilonycteris*

P. macleayi	Macleay's moustached bat	Cuba, Jamaica
P. personatus	Wagner's moustached bat	N Mexico – Brazil, Trinidad
P. quadridens (*fuliginosus*)	Sooty moustached bat	Greater Antilles; ref. 4.39

Subgenus *Phyllodia*

P. parnellii	Parnell's moustached bat	N Mexico – E Peru, Brazil, Trinidad, Greater Antilles

Mormoops; (*Aello*).

M. cuvieri (*blainvillii*)	Antillean ghost-faced bat	Greater Antilles; † C Bahamas; ref. 4.40
M. megalophylla	Peters' ghost-faced bat	S USA – Ecuador, Venezuela, Trinidad, † Cuba

Family Phyllostomidae

New-world leaf-nosed bats; *c.* 152 species; New-world tropics and subtropics; insectivorous, carnivorous, frugivorous; mainly forest

Subfamily Phyllostominae

Micronycteris; big-eared bats; Mexico – Peru, Brazil

Subgenus *Micronycteris*

M. megalotis	Brazilian big-eared bat	NE, W Mexico – E Venezuela, Peru; Trinidad, etc., Grenada
M. minuta	Gervais' big-eared bat	Nicaragua – E Peru, Bolivia, Brazil, Trinidad
M. schmidtorum	Schmidt's big-eared bat	? S Mexico, Guatemala – Colombia

Subgenus *Xenoctenes*; (possibly synonymous with *Micronycteris*); ref. 4.41.

M. hirsuta	Hairy big-eared bat	Honduras – E Peru, Brazil, Trinidad

Subgenus *Lampronycteris*

M. brachyotis	Dobson's big-eared bat	S Mexico – C Brazil; Trinidad

Subgenus *Neonycteris*

M. pusilla		E Colombia, NW Brazil

Subgenus *Trinycteris*

M. nicefori	Niceforo's big-eared bat	S Mexico – NE Peru, Brazil, Trinidad

Subgenus *Glyphonycteris*

M. behni		S Peru, C Brazil
M. sylvestris	Brown big-eared bat (Large-eared forest bat)	W, S Mexico – E Peru, SE Brazil, Trinidad

Barticonycteris; (*Micronycteris*).

B. daviesi	Davies' large-eared bat	Costa Rica, Panama, E Peru, Guyana, Surinam, Brazil; refs. 4.42, 43

Macrotus; big-eared bats; leaf-nosed bats; SW USA – Guatemala, W Indies.

M. californicus	California leaf-nosed bat	S California, S Nevada, Arizona, NW Mexico; (in *M. waterhousii*?); ref. 4.40
M. waterhousii	Waterhouse's leaf-nosed bat	N Mexico – Guatemala, Bahamas, Greater Antilles

Lonchorhina; sword-nosed bats; S Mexico – Peru, Brazil, Trinidad.

L. aurita	Tomes' long-eared bat	S Mexico – E Peru, Bolivia, Brazil, Trinidad, Bahamas (accidental; ref. 4.7)
L. fernandezi		Venezuela; ref. 4.44
L. marinkellei		E Colombia
L. orinocoensis		C Venezuela, Colombia

Macrophyllum

M. macrophyllum	Long-legged bat	S Mexico – N Argentina

Tonatia; round-eared bats; S Mexico – N Argentina.

T. bidens	Spix's round-eared bat	Guatemala – E Peru, E Brazil, Trinidad, Jamaica
T. brasiliense		E Peru, C, E Brazil; Trinidad

T. carrikeri		Venezuela, E Peru, Bolivia, Surinam
T. evotis	Davis' round-eared bat	S Mexico – Honduras
T. minuta (*nicaraguae*)	Pygmy round-eared bat	S Mexico – E Peru; (in *S. brasiliense*?); ref. 4.45
T. schultzi		Surinam; ref. 4.46
T. silvicola	D'Orbigny's round-eared bat	S Mexico – N Argentina
T. venezuelae		Venezuela; (in *T. brasiliense*?); ref. 4.45

Mimon; Gray's spear-nosed bats; S Mexico – Bolivia, Brazil.

Subgenus *Mimon*

M. bennettii	Bennett's spear-nosed bat	Guyana, Surinam, SE Brazil
M. cozumelae	Cozumel spear-nosed bat	S Mexico – N Colombia; (in *M. bennettii*?)

Subgenus *Anthorhina*

M. crenulatum (*koepckeae*)	Striped spear-nosed bat	S Mexico – Peru, Bolivia, Brazil, Trinidad

Phyllostomus; spear-nosed bats; S Mexico – Bolivia, N Argentina.

P. discolor	Pale spear-nosed bat	S Mexico – N Argentina
P. elongatus		Colombia, Venezuela – E Peru, Bolivia, SE Brazil
P. hastatus	Greater spear-nosed bat	Honduras – Peru – Paraguay; Trinidad
P. latifolius		SE Colombia, Guyana, Surinam, C Brazil

Phylloderma

P. stenops	Peters' spear-nosed bat	S Mexico – Peru, NE Brazil; Surinam, Trinidad

Trachops

T. cirrhosus	Fringe-lipped bat	S Mexico – E Peru, Bolivia, S Brazil; Trinidad

Chrotopterus

C. auritus	Peters' woolly false vampire bat	S Mexico – Paraguay, N Argentina

Vampyrum

V. spectrum	American false vampire bat (Linnaeus' false vampire bat)	S Mexico – Peru, SW Brazil, Trinidad, ? Jamaica

Subfamily Glossophaginae

Glossophaga; long-tongued bats; Mexico – N Argentina, W Indies, Bahamas.

G. commissarisi	Commissari's long-tongued bat	W Mexico – Panama; ? Colombia, N Peru
G. leachii (*alticola*) (*morenoi*)	Davis' long-tongued bat	C Mexico – Costa Rica; ref. 4.47
G. longirostris	Miller's long-tongued bat	Colombia, Venezuela, Lesser Antilles, Trinidad, etc.
G. mexicana		S Mexico; ref. 4.47
G. soricina	Pallas' long-tongued bat	N Mexico – N Argentina, Trinidad etc., Jamaica, Bahamas

Monophyllus

M. plethodon	Barbados long-tongued bat (Insular long-tongued bat)	Lesser Antilles, † Puerto Rico
M. redmani	Jamaican long-tongued bat (Leach's long-tongued bat)	Cuba, Hispaniola, Puerto Rico, Jamaica, S Bahamas

Leptonycteris; Saussure's long-tongued bats; SW USA – Venezuela.

L. curasoae		Colombia, Venezuela, , Curaçao I, etc.
L. nivalis	Big long-nosed bat	SE Arizona, W Texas – Guatemala
L. yerbabuenae (*sanborni*)	Little long-nosed bat	S Arizona, New Mexico – El Salvador

Lonchophylla

L. bokermanni		SE Brazil
L. concava	Goldman's long-tongued bat	Costa Rica, Panama, ? Peru; (in *L. mordax*?)
L. dekeyseri		Brazil; ref. 4.48
L. handleyi		Ecuador, Peru; ref. 4.49
L. hesperia		Peru, Ecuador

L. mordax	Brazilian long-tongued bat	Ecuador, Bolivia, Brazil, ?Peru
L. robusta	Panama long-tongued bat	Nicaragua – Peru
L. thomasi	Thomas's long-tongued bat	Panama, Peru, Bolivia, Brazil, Surinam

Lionycteris
L. spurrelli	Little long-tongued bat	Panama, E Peru, Ecuador, N Brazil, French Guiana

Anoura; Geoffroy's long-nosed bats; Mexico – Brazil, NW Argentina.

A. caudifera		Colombia, Venezuela – E Peru, E Brazil
A. cultrata (*brevirostrum*) (*werckleae*)		Costa Rica, Panama, Venezuela, Colombia, Peru
A. geoffroyi	Geoffroy's tail-less bat	N Mexico – NW Argentina; Trinidad, Grenada
A. latidens		Venezuela – Peru; ref. 4.50

Scleronycteris
S. ega		Venezuela, Brazil

Lichonycteris
L. degener		NE Brazil, Venezuela, Bolivia; (in *L. obscura*?); ref. 4.45
L. obscura	Brown long-nosed bat	Guatemala – Peru, Surinam

Hylonycteris
H. underwoodi	Underwood's long-tongued bat	W, S Mexico – W Panama

Platalina
P. genovensium		Peru

Choeroniscus; long-tailed bats; Mexico – Peru, Brazil.

C. godmani	Godman's bat	W Mexico – Colombia – Surinam
C. intermedius		Trinidad, E Peru, Brazil, Guianas
C. minor (*inca*)		Bolivia – Colombia – Guianas, Brazil

C. periosus		W Colombia

Choeronycteris
C. mexicana	Mexican long-tongued bat	S California, S Arizona, S New Mexico – Honduras, NW Venezuela

Musonycteris; (*Choeronycteris*).
M. harrisoni	Banana bat	W, C Mexico

Subfamily Carolliinae

Carollia; short-tailed leaf-nosed bats; Mexico – E Peru, Bolivia, S Brazil.
C. brevicauda	Silky short-tailed bat	E Mexico – E Peru, Bolivia, NE Brazil
C. castanea	Allen's short-tailed bat	Honduras – E Peru, Bolivia, Venezuela, Guianas
C. perspicillata	Seba's short-tailed bat	S Mexico – S Brazil, Paraguay; Trinidad, Tobago, Grenada, ? Jamaica
C. subrufa	Hahn's short-tailed bat	W Mexico – Nicaragua

Rhinophylla
R. alethina		W Colombia
R. fischerae		Colombia – E Peru, NW Brazil
R. pumilio		Bolivia – Colombia – French Guiana

Subfamily Sturnirinae

Sturnira; yellow-shouldered bats, American epauletted bats; Mexico – N Argentina, Uruguay, ? Chile, Trinidad, Lesser Antilles, Jamaica.

Subgenus *Sturnira*

S. aratathomasi		SW Colombia, W Ecuador
S. bogotensis		Bolivia – Venezuela
S. erythromos		Bolivia – Venezuela
S. lilium	Yellow-shouldered bat	N Mexico – N Argentina, Uruguay, ? Chile; Trinidad, Lesser Antilles, ? Jamaica

S. ludovici	Anthony's bat	N Mexico – Bolivia, Venezuela
S. luisi		Costa Rica – Peru; ref. 4.51
S. magna		Amazonian Colombia – Bolivia
S. oporaphilum		Bolivia; ref. 4.52
S. thomasi	Sofaian bat (Thomas's epauletted bat)	Guadeloupe, Lesser Antilles
S. tildae		Bolivia – Guianas, Brazil, Trinidad

Subgenus *Sturnirops*

S. mordax	Hairy-footed bat (Talamancan bat)	Costa Rica

Subgenus *Corvira*

S. bidens	E Peru – Venezuela
S. nana	S Peru

Subfamily Stenoderminae

Uroderma; tent-building bats; Mexico – S Peru, Bolivia, SE Brazil, Trinidad.

U. bilobatum	Tent-building bat (Tent-making bat)	S Mexico – S Peru, Bolivia, SE Brazil, Trinidad
U. magnirostrum	Davis' bat	S Mexico – E Peru, N Bolivia, Brazil

Vampyrops; (*Platyrrhinus*); broad-nosed bats, white-lined bats; S Mexico – N Argentina, Paraguay, Uruguay.

V. aurarius		E Venezuela – Surinam; (in *V. dorsalis*?)
V. brachycephalus		Colombia – E Peru, Guyana, Surinam
V. dorsalis	Thomas's broad-nosed bat	Panama – Bolivia, ? Venezuela
V. helleri	Heller's broad-nosed bat	S Mexico – Paraguay, Brazil, Trinidad
V. infuscus		Colombia – Bolivia, Brazil
V. lineatus (*nigellus*)		Colombia, E Peru, Surinam, C, E Brazil – Uruguay

V. recifinus		E Brazil, Guyana
V. umbratus		Colombia, Venezuela; ref.
(*oratus*)		4.53
(*aquilus*)		
V. vittatus	Greater broad-nosed bat	Costa Rica – E Peru, Venezuela

Vampyrodes

V. caraccioli	Great stripe-faced bat	Venezuela – N Brazil; Trinidad
V. major	San Pablo bat	S Mexico – Peru; ref. 4.40
(*ornatus*)		

Vampyressa; yellow-eared bats; S Mexico – E Peru, Brazil, Guyana.

Subgenus *Vampyressa*

V. melissa		E Peru
V. pusilla	Little yellow-eared bat	S Mexico – E Peru, SE Brazil

Subgenus *Metavampyressa*

V. brocki		Colombia – Surinam, Brazil
V. nymphaea	Big yellow-eared bat	Nicaragua – W Colombia

Subgenus *Vampyriscus*

V. bidens		Colombia – E Peru, N Brazil, Guyana, Surinam

Chiroderma; white-lined bats; big-eyed bats; Mexico – Peru, Bolivia, Brazil, Trinidad, Guadeloupe, Lesser Antilles.

C. doriae		E Brazil
C. improvisum	Guadeloupe white-lined bat	Guadeloupe, Montserrat, Lesser Antilles
C. salvini	Salvin's white-lined bat	W, C Mexico – Bolivia, Venezuela
C. trinitatum	Goodwin's bat	Panama, E Peru, Bolivia, Brazil, Trinidad
C. villosum	Shaggy-haired bat	S Mexico – E Peru, Bolivia, Brazil, Trinidad

Ectophylla; refs. 4.54, 55.

E. alba	White bat	Nicaragua – W Panama

Mesophylla; refs. 4.55, 56.

| *M. macconnelli* | McConnell's bat | Costa Rica – E Peru, Bolivia, Brazil, Trinidad |

Artibeus; American fruit bats; Mexico – Peru, Bolivia, Brazil, Trinidad, Bahamas, Antilles.

A. anderseni		Bolivia – Ecuador, W Brazil, French Guiana
A. aztecus	Highland fruit-eating bat	W, C Mexico – W Panama
A. cinereus	Gervais' fruit-eating bat	Colombia, Venezuela – Bolivia, E Peru, Brazil, Trinidad, Tobago, Grenada
A. concolor		Colombia, Venezuela – E Peru, Brazil, Surinam
A. fraterculus		S Ecuador, N Peru
A. fuliginosus		Peru – Colombia – Brazil, Guianas
A. glaucus		NW Colombia – C Peru; possibly referable to *A. cinereus*; ref. 4.56
A. hirsutus	Hairy fruit-eating bat	W Mexico
A. inopinatus	Honduran fruit-eating bat	El Salvador, Honduras, Nicaragua; (in *A. hirsutus*?)
A. intermedius		C, NE Mexico – Panama; ref. 4.57
A. jamaicensis	Jamaican fruit-eating bat	N Mexico – Paraguay, Brazil, Bahamas, Antilles, Trinidad
A. lituratus	Great fruit-eating bat	S Mexico – N Argentina, Lesser Antilles, Trinidad, Tobago
A. phaeotis	Dwarf fruit-eating bat (Pygmy fruit-eating bat)	W Mexico – E Peru; Guyana
A. planirostris		Guianas – Colombia – N Argentina
A. toltecus	Lowland fruit-eating bat	C Mexico – Panama
A. watsoni	Thomas's fruit-eating bat	S Mexico – Colombia; possibly referable to *A. cinereus*; ref. 4.56

Enchisthenes; possibly a subgenus of *Artibeus*.

| *E. hartii* | Little fruit-eating bat | NE Mexico – Bolivia; S Arizona, Trinidad |

Ardops
A. nichollsi   Tree bat   Lesser Antilles

Phyllops; fig-eating bats, falcate-winged bats.
P. falcatus   Cuban fig-eating bat   Cuba; † I of Pines
P. haitiensis   Dominican fig-eating bat   Hispaniola

Ariteus
A. flavescens   Jamaican fig-eating bat   Jamaica

Stenoderma
S. rufum   Red fruit bat   Puerto Rico, Virgin Is
                      (Desmarest's fig-eating
                      bat)

Pygoderma
P. bilabiatum   Ipanema bat   Surinam – Paraguay, N
                                                  Argentina

Ametrida
A. centurio   Venezuela, Guianas,
                                                  Brazil, Trinidad

Sphaeronycteris
S. toxophyllum   Colombia, E Peru –
                                                  Venezuela, Bolivia

Centurio
C. senex   Wrinkle-faced bat   W, NE Mexico – Panama;
                                                  Venezuela, Trinidad

Subfamily Phyllonycterinae

Brachyphylla; ref. 4.58.
B. cavernarum   St Vincent fruit-eating bat Puerto Rico, Lesser
                      (Antillean fruit-eating   Antilles, Virgin Is
                      bat)
B. nana   Cuban fruit-eating bat   Cuba, Grand Cayman, S
  (pumila)   Bahamas, Hispaniola,
                                                   † Jamaica

Erophylla; Greater Antilles.
E. bombifrons   Brown flower bat   Hispaniola, Puerto Rico;
                                                   refs. 4.40, 59
E. sezekorni   Buffy flower bat   Bahamas, Cuba, Jamaica,
                                                   Cayman Is

Phyllonycteris
Subgenus *Phyllonycteris*

P. major†	Puerto Rican flower bat	Puerto Rico, probably extinct
P. poeyi (*obtusa*)	Cuban flower bat	Cuba, Hispaniola

Subgenus *Reithronycteris*

P. aphylla	Jamaican flower bat	Jamaica

Subfamily Desmodontinae

Vampire bats; S USA – C Chile, N Argentina, Uruguay.

Desmodus

D. rotundus	Common vampire bat	N Mexico – C Chile, N Argentina, Uruguay, Trinidad, † Cuba

Diaemus; possibly congeneric with *Desmodus*; ref. 4.56.

D. youngi	White-winged vampire	NE Mexico – E Peru – N Argentina, Brazil; Trinidad

Diphylla

D. ecaudata	Hairy-legged vampire bat	S Texas – E Peru, S Brazil

Family Natalidae

Funnel-eared bats, long-legged bats; *c.* 4 species; Mexico – Brazil, Bahamas, Antilles.

Natalus
Subgenus *Natalus*

N. stramineus (*major*) (*espiritosantensis*)	Mexican funnel-eared bat	N Mexico – Brazil, Guianas, Antilles, † Cuba; refs. 4.60, 61
N. tumidirostris		Colombia – Surinam; Trinidad, Curacao I

Subgenus *Chilonatalus*

N. micropus (*tumidifrons*) (*macer*) (*brevimanus*)	Cuban funnel-eared bat (Jamaican long-legged bat)	Jamaica, Cuba, Bahamas, Old Providence I off Nicaragua

Subgenus *Nyctiellus*

N. lepidus	Gervais' funnel-eared bat	Bahamas, Cuba, Isle of Pines

Family Furipteridae

Smoky bats, thumbless bats; 2 species; Panama – N Chile, Brazil, Guianas, Trinidad.

Furipterus

F. horrens	Eastern smoky bat	Costa Rica – E Peru, Guianas, SE Brazil, Trinidad

Amorphochilus

A. schnablii		W Ecuador – N Chile

Family Thyropteridae

Disc-winged bats, New-world sucker-footed bats; 2 species; S Mexico – Peru, S, E Brazil.

Thyroptera

T. discifera	Peter's disk-winged bat (Honduran disk-winged bat)	Nicaragua – E Peru, Guianas, Brazil
T. tricolor	Spix's disk-winged bat	S Mexico – E Peru, Guianas, S, E Brazil, Trinidad

Family Myzopodidae

One species.

Myzopoda

M. aurita	Sucker-footed bat	Madagascar

Family Vespertilionidae

Vespertilionid bats; *c.* 347 species; cosmopolitan except some small oceanic islands, Arctic and Antarctic regions beyond limits of tree growth.

Myotis; (*Leuconoe*); mouse-eared bats, little brown bats; cosmopolitan, except some oceanic islands, Arctic, Antarctic.

M. abei		Sakhalin I

M. adversus	Large-footed bat	Taiwan, Malaya – Solomon Is, Vanuatu, N, E Australia
M. aelleni		SW Argentina
M. albescens (argentatus)	Silver-tipped myotis	S Mexico – N Argentina, Uruguay
M. altarium		S China
M. annectans	Hairy-faced bat	NE India – NE Thailand
M. atacamensis		S Peru, N Chile
M. ater		Siberut I, Borneo, Philippines, Sulawesi, ?Australia; ref. 4.12
M. auriculus	Mexican long-eared myotis	New Mexico, S Arizona – SC Mexico
M. australis	Small-footed myotis	New South Wales; (in M. ater?); ref. 4.12
M. austroriparius	South-eastern myotis	N Carolina, Indiana – Louisiana, Florida
M. bartelsi		Java
M. bechsteini	Bechstein's bat	Spain, England – W Russia, Caucasus, Iran
M. blythii	Lesser mouse-eared bat	Spain, Morocco – Afghanistan – S China
M. bocagii	Rufous mouse-eared bat	Liberia – Kenya – Angola, Transvaal; S Yemen
M. bombinus		E Asia, Japan; ref. 4.62
M. brandtii	Brandt's bat	Spain, Britain Urals
M. browni		Mindanao, Philippines; (in M. muricola ?); ref.4.12
M. californicus	California myotis	Alaska – Guatemala
M. capaccinii	Long-fingered bat	N Africa, Spain – Uzbekistan
M. carteri		W Mexico; ref. 4.63
M. chiloensis		Chile
M. chinensis	Large myotis	S China, N Thailand
M. ciliolabrum (subulatus)		SW Canada – Nebraska, Oklahoma, W Mexico; ref. 4.107
M. cobanensis	Guatemalan myotis	Guatemala; status doubtful
M. dasycneme	Pond bat	E France – Manchuria
M. daubentonii (? nathalinae) (? petax)	Daubenton's bat	Spain, Britain – E Siberia, Manchuria, Sakhalin, Hokkaido; ref. 4.64
M. dominicensis	Dominican myotis	Dominica, Lesser Antilles

M. dryas		S Andaman Is, Indian Ocean
M. elegans	Elegant myotis	Mexico – Costa Rica
M. emarginatus	Geoffroy's bat	SW Europe – Russian Turkestan, E Iran, N Africa
M. evotis	Long-eared myotis	SW Canada – NW Mexico
M. fimbriatus		Fujian, SE China
M. findleyi	Findley's myotis	Tres Marias Is, W Mexico; ref. 4.63
M. formosus	Hodgson's bat	E Afghanistan – Korea, S China, Taiwan
M. fortidens	Cinnamon myotis	W Texas – Guatemala; ref. 4.65
M. frater		Turkestan – E Siberia, SE China, Japan
M. goudoti	Malagasy mouse-eared bat	Madagascar; Anjouan I, Comoro Is
M. grisescens	Grey myotis	Oklahoma – Kentucky – Georgia; *
M. hasseltii	Lesser large-footed bat	Sri Lanka; Burma – Malaya, Java, Borneo
M. hermani		NW Sumatra
M. herrei		Luzon, Philippines
M. horsfieldii		S China – Java, Bali, Borneo, Sulawesi
M. hosonoi		N, C Honshu, Japan
M. ikonnikovi		E Siberia, N Korea, Sakhalin, Hokkaido
M. insularum		? Samoa
M. jeannei		Mindanao, Philippines
M. keaysi	Hairy-legged myotis	NE Mexico – Venezuela – Peru; Trinidad
M. keenii (*septentrionalis*)	Keen's myotis	Alaska – Washington, Manitoba – Newfoundland – Florida
M. leibii	Least brown bat (Small-footed myotis)	SE Canada – Oklahoma, Georgia; ref. 4.107
M. lesueuri	Lesueur's hairy bat	Cape Prov., S Africa; (in *M. seabrai*?); ref. 4.66
M. levis		S Brazil – Paraguay, Uruguay, Argentina
M. longipes		Afghanistan, Kashmir, ? Vietnam

M. lucifugus (occultus)	Little brown myotis	Alaska, SE Canada – C Mexico
M. macrodactylus		E Siberia, S Kurile Is, Japan
M. macrotarsus		Philippines, Borneo
M. martiniquensis	Schwartz's myotis	Martinique, Barbados, Lesser Antilles
M. milleri	Miller's myotis	Baja California, Mexico
M. montivagus	Burmese whiskered bat	S India, Burma, S China, Malaya, Borneo
M. morrisi		Ethiopia
M. muricola		Afghanistan – Java – New Guinea, ? Philippines, Taiwan
M. myotis	Large mouse-eared bat	SW Europe – Asia Minor; Azores; ref. 4.67
M. mystacinus	Whiskered bat	Ireland – Japan, N Iran, N India, Tibet, Morocco
M. nattereri	Natterer's bat	Morocco, W Europe – Urals – Israel; ref. 4.62
M. nesopulos (larensis)		NW Venezuela, Curaçao I, Bonaire I; ref. 4.68
M. nigricans	Black myotis	W, NE Mexico – N Argentina, Trinidad, Tobago, Grenada
M. oreias	Singapore whiskered bat	Malaya
M. oxyotus	Montane myotis	Costa Rica – Peru, N Bolivia
M. ozensis		C Honshu, Japan; status doubtful
M. patriciae		Mindanao, Philippines
M. peninsularis	Peninsular myotis	Baja California
M. pequinius		E China
M. peshwa		India
M. planiceps	Flat-headed myotis	N Mexico
M. pruinosus		N Honshu, Japan
M. ricketti	Rickett's big-footed bat	E China
M. ridleyi	Ridley's bat	Malaya, ? Sumatra; Borneo
M. riparius	Riparian myotis	Honduras – Peru – Uruguay; Trinidad
M. rosseti	Thick-thumbed myotis	Thailand, Kampuchea
M. ruber		SE Brazil, Paraguay, N Argentina

M. rufopictus		Philippines
M. schaubi		Armenia, W Iran; ref. 4.62
(*araxenus*)		
M. scotti	Scott's mouse-eared bat	Ethiopia
M. seabrae	Angola wing-gland bat	Angola – Cape Prov.
	(Angola hairy bat)	
M. sicarius		Nepal, Sikkim
M. siligorensis	Himalayan whiskered bat	N India – S China – Malaya, Borneo; ref. 4.69
M. simus		Panama – Paraguay, Brazil
M. sodalis	Social bat	C, E USA; *
	(Indiana myotis)	
M. stalkeri	Kei myotis	Kei Is, New Guinea
M. surinamensis		? Surinam; status doubtful
M. thysanodes	Fringed myotis	SW Canada – S Mexico
M. tricolor	Cape hairy bat	Ethiopia – Zaire, S Africa
M. velifer	Cave myotis	S USA – Honduras
M. volans	Long-legged myotis	Alaska – S Mexico
	(Hairy-winged bat)	
M. weberi	Orange-winged myotis	S Sulawesi
M. welwitschii	Welwitsch's hairy bat	Ethiopia – Zaire, Mozambique, S Africa
M. yesoensis		Hokkaido, Japan; ref. 4.70
M. yumanensis	Yuma myotis	British Columbia – C USA – C Mexico

Pizonyx; (*Myotis*).

P. vivesi	Fish-eating bat	NW Mexico; coasts, islands

Lasionycteris

L. noctivagans	Silver-haired bat	Alaska, S Canada – NE Mexico; Bermuda

Eudiscopus

E. denticulus	Disc-footed bat	Burma, Laos

Pipistrellus; (*Vespertilio, Perimyotis*, ref. 4.71); pipistrelles; Europe, Africa – Solomon Is, Australia, Tasmania; S Canada – Honduras.

P. aero		NW Kenya, ? Ethiopia
P. affinis	Chocolate bat	NE Burma, Yunnan, India
P. anchietae	Anchieta's pipistrelle	Angola, S Zaire, Zambia

P. angulatus (*collinus*) (*ponceleti*)	Greater New Guinea pipistrelle	New Guinea, Bismarck Arch., Solomons; refs 4.12, 72
P. anthonyi		N Burma
P. arabicus		N Oman; ref. 4.73
P. ariel	Desert pipistrelle	Egypt, Sudan
P. babu		Afghanistan, N, C India – Burma, China
P. bodenheimeri	Bodenheimer's pipistrelle	Israel, SW Arabia, Oman, ? Socotra I
P. cadornae	Cadorna's pipistrelle	NE India – Thailand
P. ceylonicus	Kelaart's pipistrelle	Pakistan – S China, Vietnam, Borneo, Sri Lanka
P. circumdatus	Gilded black pipistrelle	N Burma – Java
P. coromandra	Indian pipistrelle	E Afghanistan – S China, Vietnam, Sri Lanka, Nicobar Is
P. crassulus	Broad-headed pipistrelle	Cameroun, C Zaire
P. cuprosus		Borneo; ref. 4.14
P. curtatus		Enggano I, Sumatra, ref. 4.12
P. deserti	Desert pipistrelle	Algeria – Egypt, N Sudan, Burkina Faso
P. eisentrauti	Eisentraut's pipistrelle	Ivory Coast, Cameroun, Kenya
P. endoi		Honshu, Japan
P. hesperus	Western pipistrelle (Canyon bat)	Washington – C Mexico
P. imbricatus	Brown pipistrelle	Java, Philippines, ? Sulawesi; refs. 4.12, 72
P. inexspectatus	Aellen's pipistrelle	Benin – Uganda, ? Sudan; ref. 4.74
P. javanicus	Javan pipistrelle	Japan, E Siberia – Java, Borneo, Sulawesi, Philippines, ? N Australia
P. joffrei	Joffre's pipistrelle	Burma
P. kitcheneri		Borneo
P. kuhli	Kuhl's pipistrelle	Africa, SW Europe – Kashmir
P. lophurus		S Thailand, S Burma
P. macrotis		Malaya, Sumatra, Bali; ref. 4.14

P. maderensis	Madeira pipistrelle	Madeira; Canary Is
P. mimus	Indian pygmy pipistrelle	Afghanistan – Vietnam, Sri Lanka
P. minahassae	Minahassa pipistrelle	N Sulawesi
P. mordax		Java
P. murrayi		Christmas I, Indian Ocean; ? Cocos-Keeling Is; possibly subspecies of *P. tenuis*; ref. 4.72
P. musciculus	Least pipistrelle	Cameroun, C Zaire, Gabon
P. nanulus	Tiny pipistrelle	Sierra Leone – Uganda; ref. 4.74
P. nanus	Banana bat	Sierra Leone – Somalia – S Africa, Madagascar
P. nathusii	Nathusius' pipistrelle	Spain – Urals, Caucasus
P. peguensis		Pegu, S Burma
P. permixtus	Dar-es-Salaam pipistrelle	Tanzania
P. petersi	Peters' pipistrelle	N Sulawesi, Buru, Amboina, S Moluccas
P. pipistrellus	Common pipistrelle	W Europe, Morocco – Kashmir, ? Korea, Japan, Taiwan
P. pulveratus	Chinese pipistrelle	S China, Thailand
P. rueppelli	Rüppell's bat	Senegal – Tanzania – Botswana, Egypt, Iraq
P. rusticus	Rusty bat	Ghana – Ethiopia – Namibia, Transvaal
P. savii	Savi's pipistrelle	Canary, Cape Verde Is, Iberia, Morocco – Korea, Japan, Burma
P. societatis		Malaya
P. stenopterus		Malaya, Sumatra, Borneo, Philippines
P. sturdeei		Bonin Is, S of Japan
P. subflavus	Eastern pipistrelle	SE Canada – Honduras
P. tasmaniensis	Great pipistrelle	SW, E Australia, Tasmania
P. tenuis (*papuanus*)	Least pipistrelle	S Thailand – Java, Borneo, Philippines, Bismarck Arch., Sulawesi, Timor, N Australia; refs. 4.12, 72
P. vordermanni		Borneo, Billiton I; ref. 4.12

Scotozous
S. dormeri   India, Pakistan

Nyctalus
N. aviator   Korea, Japan, E China
N. azoreum   Azores Is; ref. 4.75
N. lasiopterus   Giant noctule   SW Europe – Iran; N
                                                      Africa
N. leisleri   Lesser noctule   Madeira, N Africa, W
                          (Leisler's bat)   Europe – N India
N. montanus   E Afghanistan – N India
N. noctula   Noctule   Britain, Algeria, W Europe
                                                      – China, Japan, Taiwan,
                                                      Vietnam, ? Malaya; ref.
                                                      4.76

Glischropus
G. javanus   Java
G. tylopus   Thick-thumbed pipistrelle   Burma – Sumatra, Borneo,
                                                      Palawan, N Moluccas

Eptesicus; (*Vespertilio*); serotines, brown bats; Europe – E Asia, Africa, Australia;
Alaska – Argentina.

Subgenus *Eptesicus*
E. bobrinskoi   Kazakhstan – NW Iran
E. bottae   Botta's serotine   NE Egypt – Arabia –
                                                      Turkestan, Iran
E. brasiliensis   Brazilian brown bat   C Mexico – Uruguay,
  (andinus)   SE Argentina; Trinidad
E. brunneus   Dark brown serotine   Ivory Coast – C Zaire
E. capensis   Cape serotine   Africa S of Sahara,
                                                      Ethiopia, Madagascar
E. chiriquinus   Chiriqui brown bat   Panama; (in E.
                                                      brasiliensis?); ref. 4.56
E. demissus   Surat serotine   S Thailand
E. diminutus   E, SE Brazil –
                                                      N Argentina, Uruguay
E. douglasorum   Yellow-lipped bat   NW Australia
E. flavescens   Yellow serotine   Angola, Burundi; ref. 4.77
E. furinalis   Argentine brown bat   Mexico – N Argentina;
  (montosus)   ref. 4.56

E. fuscus	Big brown bat	Alaska, S Canada – Colombia, Venezuela, Greater Antilles, Dominica; ref. 4.78
E. guadeloupensis	Guadeloupe brown bat	Guadeloupe, Lesser Antilles
E. guineensis	Tiny serotine	Senegal – Ethiopia, NE Zaire
E. hottentotus	Long-tailed house bat	Namibia – Mozambique, S Africa
E. innoxius		W Ecuador, W Peru
E. kobayashii		Korea; status uncertain
E. loveni	Loven's serotine	W Kenya
E. lynni	Lynn's brown bat	Jamaica
E. melckorum	Melck's house bat	Zambia, Tanzania, Mozambique, SW Cape Prov.
E. nasutus (*walli*)	Sind bat	S Arabia – Pakistan; ref. 4.79
E. nilssonii	Northern bat	France, Norway – E. Siberia – Nepal, Japan, Iraq
E. pachyotis	Thick-eared bat	Assam – N Thailand
E. platyops	Lagos serotine	Senegal, Nigeria; possibly subspecies of *E. serotinus*
E. pumilus	Little bat	N, C, E Australia
E. regulus	King River bat	SW, SE Australia
E. rendalli	Rendall's serotine	Gambia – Somalia – Mozambique, Botswana
E. sagittula	Large forest bat	NE, SE Australia, Lord Howe I, Tasmania
E. serotinus	Serotine	England, Morocco, W Europe – Thailand, China, Korea
E. somalicus	Somali serotine	Guinea-Bissau – Somalia – Kenya, Tanzania, ? Namibia
E. tatei		NE India; status uncertain
E. tenuipinnis	White-winged serotine	Guinea – Kenya – Angola
E. vulturnus	Little forest bat	C, S, SE Australia, Tasmania
E. zuluensis	Aloe serotine	Namibia, Zambia – S Africa; (in *E. somalicus*?)

Subgenus *Rhinopterus*

E. floweri	Horn-skinned bat	Mali, S Sudan

Ia

I. io	Great evening bat	Assam – S China, Indochina, Thailand

Vespertilio

V. murinus	Particolored bat	Scandinavia, Siberia – Iran, Afghanistan
V. orientalis		E China, Honshu, Taiwan
V. superans		E Siberia, E China, Japan

Laephotis; Africa S of Sahara.

L. angolensis		Angola, S Zaire
L. botswanae	Botswana long-eared bat	S Zaire, Zambia, NW Botswana
L. namibensis	Namib long-eared bat	Namibia, SW Cape Prov.
L. wintoni	De Winton's long-eared bat	Ethiopia, Kenya

Histiotus; big-eared brown bats; Colombia – Chile, Argentina.

H. alienus		Brazil, Uruguay
H. laephotis		Bolivia
H. macrotus		NW Argentina, Chile, Peru
H. montanus		Venezuela, Colombia – Chile, Argentina
H. velatus		Brazil, Paraguay

Philetor

P. brachypterus	New Guinea brown bat	E Nepal, Malaya – New Guinea, Bismarck Arch., Philippines; ? Java; refs. 4.80, 81

Tylonycteris

T. pachypus	Bamboo bat (Lesser club-footed bat) (Flat-headed bat)	India, S China – Java, Lesser Sundas, Borneo, Philippines, Andaman Is
T. robustula	Greater club-footed bat (Flat-headed bat)	SW China – Java, Borneo, Sulawesi, Timor

Mimetillus

M. moloneyi	Moloney's flat-headed bat	Sierra Leone – Ethiopia – Angola

Hesperoptenus
Subgenus *Hesperoptenus*

H. doriae	False serotine bat	Malaya, Borneo

Subgenus *Milithronycteris*

H. blanfordi	Blanford's bat	S Burma – Malaya, Borneo
H. gaskelli		Sulawesi; ref. 4.12
H. tickelli	Tickell's bat	India – Thailand, Andaman Is, Sri Lanka, ? S China
H. tomesi		Malaya, Borneo

Glauconycteris; Africa S of Sahara.

G. alboguttatus	Allen's striped bat	E Zaire, Cameroun
G. argentata	Silvered bat	Cameroun – Kenya – NE Angola, Tanzania
G. beatrix	Beatrix bat	Ivory Coast – Kenya
G. egeria	Bibundi bat	Cameroun, Uganda
G. gleni		Cameroun, Uganda
G. kenyacola		E Kenya; ref. 4.82
G. machadoi	Machado's butterfly bat	E Angola; possibly subspecies of *G. variegata*; ref. 4.83
G. poensis	Abo bat	Senegal – Uganda; ref. 4.74
G. superba	Pied bat	Ivory Coast, Ghana, NE Zaire
G. variegata	Butterfly bat	Senegal – Somalia – S Africa

Chalinolobus; Australia.

C. dwyeri	Large pied bat	S Queensland – E New South Wales
C. gouldii	Gould's wattled bat	Australia, New Caledonia
C. morio	Chocolate wattled bat	S Australia, Tasmania
C. nigrogriseus	Hoary bat (Frosted bat)	SE New Guinea, Fergusson I, N Australia
C. picatus	Little pied bat	C, S Queensland – Victoria
C. tuberculatus	Long-tailed bat	New Zealand

Scotoecus; Africa – N India.

S. albofuscus	Light-winged lesser house bat (Thomas's house bat)	Senegal – Tanzania, S Malawi, Mozambique, ? Zambia

S. hindei		Nigeria – Somalia – Tanzania – Angola
S. hirundo		Senegal – Ethiopia; refs. 4.74, 84
S. pallidus		Pakistan, N India

Nycticeius

N. humeralis (*cubanus*)	Evening bat (Twilight bat)	C, SE USA – E Mexico, Cuba
N. schlieffenii	Schlieffen's bat	Mauretania – Egypt – Namibia, Mozambique, SW Arabia

Scoteanax; (*Nycticeius*); ref. 4.85.

S. rueppellii	Rüppell's broad-nosed bat (Greater broad-nosed bat)	E Queensland, E New South Wales

Scotorepens; (*Nycticeius*); ref. 4.85.

S. balstoni (*influatus*)	Western broad-nosed bat	NW, N, C Australia, New Guinea
S. greyi (*caprenus*) (*aquilo*)	Little broad-nosed bat (Grey's bat)	Australia except SW, NE, SE
S. orion		SE Australia
S. sanborni	Little northern broad-nosed bat	NE Australia

Rhogeessa; C, S America.

R. genowaysi		Chiapas, Mexico; ref. 4.86
R. gracilis	Slender yellow bat	W Mexico
R. minutilla		N Venezuela, NE Colombia
R. mira	Least yellow bat	Michoacan, C Mexico
R. parvula	Little yellow bat	W Mexico
R. tumida	Central American yellow bat	E Mexico – Ecuador, S Brazil, Bolivia; Trinidad

Baeodon; (*Rhogeessa*).

B. alleni	Allen's baeodon (Allen's yellow bat)	W, C Mexico

Scotomanes

S. emarginatus		India
S. ornatus	Harlequin bat	N India – S China, Vietnam

Scotophilus; Africa, SE Asia; classification in Africa uncertain.

S. borbonicus		Reunion, Madagascar; refs. 4.17, 87
S. celebensis	Sulawesi yellow bat	N Sulawesi
S. dinganii	African yellow house bat	Senegal – Ethiopia – S Africa
S. heathii	Asiatic greater yellow house bat	Afghanistan – S China, Vietnam, Sri Lanka
S. kuhlii	Asiatic lesser yellow house bat	Pakistan – Hainan – Timor, Aru Is, Philippines, Taiwan
S. leucogaster		Mauritania – Somalia, Aden; ? Namibia, Botswana; ref. 4.108
S. nigrita (gigas)	Greater brown bat (Giant yellow bat)	Senegal – S Sudan – Mozambique
S. nigritellus		Mali, Ivory Coast – Cent. African Rep.; (in S. borbonicus?); ref. 4.108
S. nucella		Ghana, Uganda; ref. 4.109
S. nux		Sierra Leone – Kenya
S. robustus		Madagascar; (in S. dinganii?)
S. viridis	Lesser yellow house bat	Tanzania, Angola, S Africa; (in S. borbonicus?); refs. 4.87, 109

Otonycteris

O. hemprichii	Hemprich's long-eared bat	Algeria, Niger – Egypt – Afghanistan, Kashmir

Lasiurus; (Nycteris; ref. 4.40); hairy-tailed bats.

L. borealis (pfeifferi) (degelidus) (minor) (brachyotis)	Red bat	S Canada – C Chile, Argentina; Bahamas, Trinidad, Greater Antilles, Puerto Rico, Bermuda, Galapagos; ref. 4.40

L. castaneus	Tacarcuna bat	Panama
L. cinereus	Hoary bat	NC, S Canada – C Chile, C Argentina; Hawaii, Galapagos; vagrant Iceland, Orkney, Bermuda; (*)
L. seminolus	Seminole bat	E USA

Dasypterus; (*Lasiurus*).

D. ega	Southern yellow bat	SW USA – Argentina
D. egregius		Panama, Brazil
D. insularis		Cuba; refs. 4.39, 40, 88
D. intermedius	Northern yellow bat (Eastern yellow bat)	New Jersey, N Mexico – Honduras

Barbastella; barbastelles.

B. barbastellus	Western barbastelle	England, France, Morocco – Caucasus
B. leucomelas	Eastern barbastelle	Caucasus – N India, W China, Japan, ? NE Africa

Plecotus; long-eared bats.

Subgenus *Plecotus*

P. auritus	Brown long-eared bat (Common long-eared bat)	Britain, France – NE China, Korea, Japan, ? N India
P. austriacus	Grey long-eared bat	Spain, S England – W China; Cape Verde Is, Canary Is, N Africa, Senegal

Subgenus *Corynorhinus*

P. mexicanus	Mexican big-eared bat	NW, NE, C Mexico
P. rafinesquii	Rafinesque's big-eared bat (Eastern lump-nosed bat)	SE USA
P. townsendii	Townsend's big-eared bat (Lump-nosed bat)	SW Canada, W USA – Mexico; (*)

Idionycteris; (*Plecotus*).

I. phyllotis	Allen's big-eared bat	Nevada, Utah – W New Mexico – C Mexico

Euderma

E. maculatum	Spotted bat (Pinto bat)	W, SW USA – N, C Mexico; *

Subfamily Miniopterinae

Miniopterus; long-fingered bats, bent-winged bats; refs. 4.10, 12, 89, 90, 91.

M. australis (*solomonensis*)	Little long-fingerd bat	Java, Borneo, Philippines – E Australia, New Caledonia, Loyalty Is; ref. 4.91, 92
M. fraterculus	Lesser long-fingered bat	Malawi – S Africa; ? Zambia
M. fuscus (*yayeyamae*)		Ryukyu Is
M. inflatus	Greater long-fingered bat	Cameroun – Somalia, Zambia, Mozambique
M. magnater (*macrodens*) (*? bismarckensis*)		S China, Malaya – Java, Borneo, New Guinea; ref. 4.91
M. medius	SE Asian long-fingered bat	S Thailand – Java, New Guinea
M. minor	Least long-fingered bat	Congo Rep. – Tanzania, Madagascar, Comoro Is
M. pusillus (*macrocneme*)		Nicobar Is, Thailand, Philippines, Sulawesi, Java – New Caledonia, Loyalty Is
M. robustior		Loyalty Is
M. schreibersi (*oceanensis*)	Schreiber's long-fingered bat	Africa, Madagascar, SW Europe – China, Japan, Philippines, Solomons, NW, N, E Australia; ref. 4.91
M. tristis (*propitristis*) (*celebensis*) (*insularis*) (*melanesiensis*)		Sulawesi, Philippines, New Guinea – Solomons, New Hebrides; refs. 4.90, 91

Subfamily Murininae

Murina; tube-nosed bats.

Subgenus *Murina*

M. aenea	Bronze tube-nosed bat	Malaya, Borneo; ref. 4.14
M. aurata	Little tube-nosed bat	Nepal – China, Thailand, Korea, Sakhalin, Japan; ref. 4.93
M. balstoni		Java
M. canescens		Nias I (W Sumatra); (in *M. suilla?*)
M. cyclotis	Round-eared tube-nosed bat	N India – S China – Malaya, Hainan, Sri Lanka, Philippines, Borneo; ref. 4.12
M. florium	Flores tube-nosed bat	Lesser Sunda Is, ? New Guinea; S Moluccas, N Australia; ref. 4.12
M. huttoni	Hutton's tube-nosed bat	Himalayas – S China, Malaya
M. leucogaster (*fusca*)	Greater tube-nosed bat	NE India, S China – E Siberia, Japan
M. puta		Taiwan; status uncertain
M. rozendaali		N Borneo; ref. 4.14
M. silvatica		Japan; ref. 4.94
M. suilla	Brown tube-nosed bat	Thailand – Java, Borneo
M. tenebrosa		Tsushima I, Japan
M. tubinaris		N Pakistan – Vietnam

Subgenus *Harpiola*

M. grisea	Peters' tube-nosed bat	NW India

Harpiocephalus

H. harpia	Hairy-winged bat	India – Vietnam, Taiwan, Sumatra, Java, Borneo, S Moluccas
H. mordax		Burma, ? Borneo; ref. 4.14

Subfamily Kerivoulinae

Kerivoula; woolly bats.

K. africana	Tanzanian woolly bat	Tanzania

K. agnella	Louisiade trumpet-eared bat	Sudest I, SE New Guinea; St Aignan's I (= Misima I), Louisiade Arch.
K. argentata	Damara woolly bat	S Kenya – Namibia – Natal
K. cuprosa	Copper woolly bat	Ghana, S Cameroun, Kenya, ? Zaire
K. eriophora		Ethiopia; possibly conspecific with *K. africana*
K. hardwickii	Hardwick's forest bat	Sri Lanka, India – S China – Java, Lesser Sundas, Philippines, Sulawesi
K. intermedia		Borneo, Malaya; ref. 4.14
K. lanosa	Lesser woolly bat	Liberia – Ethiopia – S Africa
K. minuta	Least forest bat	S Thailand, Malaya, Borneo; ref. 4.14
K. muscina	Fly River trumpet-eared bat	C New Guinea
K. myrella	Bismarck trumpet-eared bat	Admiralty Is, Bismarck Arch.
K. papillosa	Papillose bat	NE India – Java, Borneo, Sulawesi; ref. 4.12
K. pellucida	Clear-winged bat	Philippines, Malaya, Borneo, ? Sumatra, Java
K. phalaena	Spurrell's woolly bat	Liberia, Ghana, Cameroun, Zaire
K. picta	Painted bat	Sri Lanka, S India – S China – Java, Lesser Sundas, Borneo, Ternate I
K. smithii	Smith's woolly bat	Nigeria – E Zaire, Kenya
K. whiteheadi		S Thailand, Malaya, Borneo, Philippines

Phoniscus

P. aerosa	Dubious trumpet-eared bat	? SE Asia; origin and status uncertain
P. atrox	Groove-toothed bat	S Thailand, Malaya, Sumatra, Borneo; ref. 4.14

P. jagorii	Peters' trumpet-eared bat	Samar I, Philippines, Borneo, Java, ? Sulawesi
P. papuensis	Papuan trumpet-eared bat	SE New Guinea, E Australia

Subfamily Nyctophilinae

Antrozous; (*Bauerus*); possibly a member of the Vespertilioninae.

A. koopmani	Cuban bat	Cuba
A. pallidus	Pallid bat	Br. Colombia, W, C USA – C Mexico

Bauerus; (*Antrozous*); possibly a member of the Vespertilionidae; ref. 4.95.

B. dubiaquercus (*meyeri*)	Van Gelder's bat	Tres Marias Is, Veracruz, Mexico, Honduras

Nyctophilus; (*Lamingtonia*); ref. 4.96.

N. arnhemensis	Arnhem Land long-eared bat	NW Australia
N. bifax	Northern long-eared bat	N Australia
N. geoffroyi	Lesser long-eared bat	Australia, Tasmania
N. gouldi	Gould's long-eared bat	SW, SE Australia, Tasmania
N. microdon	Small-toothed long-eared bat	SE New Guinea
N. microtis (*lophorhina*)	Papuan long-eared bat	SE New Guinea; ref. 4.96
N. timoriensis	Greater long-eared bat	SW, S, SE Australia, New Guinea; ? Tasmania, ? Timor; ref. 4.97
N. walkeri	Pygmy long-eared bat	NW Australia

Pharotis

P. imogene	Big-eared bat	SE New Guinea

Subfamily Tomopeatinae

Tomopeas

T. ravus		NW Peru

Family Mystacinidae

New Zealand short-tailed bats; 2 species.

Mystacina; New Zealand; refs. 4.98, 99.

M. robusta	New Zealand greater short-tailed bat	Solomon I, Big South Cape I; probably extinct
M. tuberculata	New Zealand lesser short-tailed bat	New Zealand; Solomon I, Big South Cape I, Codfish I, Jacky Lee I

Family Molossidae

Free-tailed bats; *c.* 90 species; tropics and subtropics of Old and New Worlds; generic classification unstable; refs. 4.100, 101.

Tadarida; free-tailed bats, mastiff bats; refs. 4.100, 101.

Subgenus *Tadarida*; (*Rhizomops*).

T. aegyptiaca	Egyptian free-tailed bat	Africa, Arabia, Iran – India, Sri Lanka
T. africana (*ventralis*)	African giant free-tailed bat	Sudan, Ethiopia – Mozambique, Transvaal
T. australis	Southern mastiff-bat (white-striped mastiff-bat)	C, S Australia
T. brasiliensis	Brazilian free-tailed bat	W, S USA – C Chile, Argentina, Bahamas, Antilles
T. espiritosantensis		Brazil; status uncertain
T. fulminans	Madagascar large free-tailed bat	E Zaire, Kenya – Zimbabwe, Madagascar
T. kuboriensis	Small-eared mastiff bat	SE New Guinea; (in *T. australis*?); ref. 4.100
T. lobata	Kenya big-eared free-tailed bat	Kenya, Zimbabwe
T. teniotis	European free-tailed bat	S Europe, Madeira, Canary Is, N Africa – N India, China, Korea, Japan, Taiwan

Subgenus *Nyctinomops*

T. aurispinosa (*similis*)	Peale's free-tailed bat	E, W Mexico, Colombia, Peru, E Brazil
T. femorosacca	Pocketed free-tailed bat	SW USA – S Mexico

T. laticaudata (europs) (gracilis) (yucatanica)	Broad-tailed bat	NE Mexico – Venezuela – Paraguay, Cuba, Trinidad
T. macrotis	Big free-tailed bat	SW British Colombia, E, C USA – Brazil, Uruguay, Paraguay, Greater Antilles

Subgenus Mops

T. condylura	Angola free-tailed bat	Gambia – Somalia – S Africa; Madagascar
T. congica	Medje free-tailed bat	Ghana, Cameroun, NE Zaire, Uganda
T. demonstrator	Mongalla free-tailed bat	Burkina Faso, Sudan, NE Zaire, Uganda
T. lanei		Mindanao, Philippines; (in T. sarasinorum?); ref. 4.100
T. midas	Midas free-tailed bat	Senegal – Ethiopia – S Africa; SW Arabia
T. mops	Malayan free-tailed bat	Malaya, Sumatra, Borneo, ? Java
T. niangarae	Niangara free-tailed bat	NE Zaire
T. niveiventer	White-bellied free-tailed bat	Zaire – Angola, Zambia, Mozambique
T. sarasinorum	Sulawesi mastiff-bat	C Sulawesi
T. trevori		S Sudan, NE Zaire, Uganda

Subgenus Xiphonycteris; ref. 4.102.

T. brachyptera		Uganda – Tanzania, Mozambique, Zanzibar
T. leonis	Sierra Leone free-tailed bat	Sierra Leone – E Zaire, Uganda, Bioco; (in T. brachyptera?); ref. 4.102
T. nanula	Dwarf free-tailed bat	Sierra Leone – Ethiopia, Kenya, Zaire
T. petersoni	Peterson's free-tailed bat	Ghana, Cameroun, ref. 4.102
T. spurrelli	Spurrell's free-tailed bat	Ivory Coast – Rio Muni, Central African Rep. Zaire, Bioco

T. thersites	Railer bat	Sierra Leone – S, E Zaire, Bioco, Mozambique, ? Zanzibar

Subgenus *Chaerephon*

T. aloysiisabaudiae	Duke of Abruzzi's free-tailed bat	Ghana, Gabon, N Zaire, Uganda, ? Ethiopia
T. ansorgei	Ansorge's free-tailed bat	Cameroun, Ethiopia – Angola
T. bemmelini	Gland-tailed free-tailed bat	Liberia – S Sudan – N Tanzania
T. bivittata	Spotted free-tailed bat	Ethiopia – Zambia – Mozambique; (in *T. ansorgei*?)
T. chapini	Chapin's free-tailed bat	W, NE Zaire, Uganda – Namibia, ? Ethiopia
T. gallagheri	Gallagher's free-tailed bat	C Zaire
T. jobensis	Northern mastiff-bat	New Guinea, N Australia, Solomons, Fiji, Vanuatu
T. johorensis	Dato Meldrum's bat	Malaya, Sumatra
T. major	Lappet-eared free-tailed bat	Ghana, Mali – S Sudan – Tanzania
T. nigeriae	Nigerian free-tailed bat	Ghana – Ethiopia, Zimbabwe, Namibia; SW Arabia
T. plicata	Wrinkle-lipped free-tailed bat	Sri Lanka, India – S China – Java, Borneo, Cocos-Keeling Is, Philippines, Hainan
T. pumila	Little free-tailed bat	Senegal – Somalia – Angola, Natal, Madagascar, SW Arabia
T. pusillus		Aldabra I, Indian Ocean; (in *T. pumila*?)
T. russata	Russet free-tailed bat	Ghana, Cameroun, NE Zaire

Mormopterus; (*Tadarida*, *Micronomus*); refs. 4.100, 101.

M. acetabulosus	Natal wrinkle-lipped bat (Natal free-tailed bat)	Ethiopia, Natal, Madagascar, Mauritius, Reunion
M. beccarii	Beccari's mastiff-bat	Amboina, New Guinea, Queensland

M. doriae		Sumatra
M. jugularis	Peters' wrinkle-lipped bat	Madagascar
M. kalinowskii		Peru, N Chile
M. loriae	Little northern mastiff-bat	New Guinea, N Australia; (in *M. planiceps*?)
M. minutus	Little goblin bat	Cuba
M. norfolkensis	Eastern little mastiff-bat (Norfolk island bat)	SE Queensland, E New South Wales, Norfolk I
M. phrudus		Peru
M. planiceps	Litle mastiff bat	SW, C Australia

Sauromys; possibly a subgenus of *Mormopterus*; refs. 4.100, 101.

S. petrophilus	Robert's flat-headed bat (Flat-headed free-tailed bat)	Namibia – Zimbabwe, Mozambique, S Africa

Platymops; possibly a subgenus of *Mormopterus*; refs. 4.100, 101.

P. setiger	Peters' flat-headed bat	SE Sudan, S Ethiopia, Kenya

Otomops

O. formosus	Java mastiff-bat	Java
O. martiensseni	Martienssen's free-tailed bat (Giant mastiff-bat) (Large-eared free-tailed bat)	Ethiopia, Cent. African Rep. – Angola, Natal, Madagascar
O. papuensis	Big-eared mastiff-bat	SE New Guinea
O. secundus	Mantled mastiff-bat	NE New Guinea
O. wroughtoni	Wroughton's free-tailed bat	S India

Myopterus

M. albatus	Banded free-tailed bat	Ivory Coast, NE Zaire
M. daubentonii	Daubenton's free-tailed bat	Senegal; status uncertain
M. whitleyi	Bini free-tailed bat	Ghana – Zaire, Uganda

Cabreramops; (*Molossops*); ref. 4.103.

C. aequatorianus	W Ecuador

Molossops

Subgenus *Molossops*

M. neglectus	Surinam; ref. 4.104
M. temmincki	Colombia – N Argentina, Uruguay

Subgenus *Cynomops*

M. abrasus (*brachymeles*)		Guianas – Venezuela – Peru – N Argentina; ref. 4.105
M. greenhalli	Greenhall's dog-faced bat	W Mexico – Ecuador, NE Brazil, Trinidad; ref. 4.106
M. milleri		Peru; (in *M. planirostris*?); ref. 4.56
M. paranus		C Mexico, Venezuela, Brazil; (in *M.* *planirostris*?); refs. 4.53, 56, 104
M. planirostris	Southern dog-faced bat	Panama – Peru – N Argentina

Neoplatymops; possibly a subgenus of *Molossops*; ref. 4.100.

N. mattogrossensis		C Venezuela, S Guyana, NE, C Brazil

Eumops

E. auripendulus	Slouch-eared bat	S Mexico – N Argentina, Trinidad, Jamaica
E. bonariensis (*nanus*)	Peters' mastiff-bat	S Mexico – C Argentina
E. dabbenei		N Venezuela, N Colombia, N Argentina, Paraguay
E. glaucinus	Wagner's mastiff-bat	S Florida; C Mexico – N Argentina – SE Brazil; Cuba, Jamaica
E. hansae	Sanborn's mastiff-bat	Costa Rica – Brazil, Guyana
E. maurus	Guianian mastiff-bat	Guyana, Surinam
E. perotis (*trumbulli*)	Greater mastiff-bat	SW USA – CW Mexico, Venezuela – N Argentina; Cuba
E. underwoodi	Underwood's mastiff-bat	S Arizona – Nicaragua; ref. 4.65

Promops

P. centralis (*davisoni*) (*occultus*)	Thomas' mastiff-bat	W Mexico – Peru – Paraguay; Trinidad
P. nasutus (*pamana*)		N Argentina – Venezuela; Trinidad; ref. 4.100

Molossus

M. ater	Black mastiff-bat	N Mexico – Guyana – Peru, N Argentina, Trinidad
M. barnesi		French Guiana, Brazil
M. bondae	Bonda mastiff-bat	Honduras – Ecuador, NW Venezuela
M. molossus (*coibensis*) (*espiritosantensis*)	Pallas' mastiff-bat	N Mexico – N Argentina, Trinidad, Antilles; refs. 4.40, 60, 61
M. pretiosus (*macdougalli*)	Miller's mastiff-bat	S Mexico – Venezuela; ref. 4.40
M. sinaloae	Allen's mastiff-bat	W Mexico – Costa Rica
M. trinitatis	Trinidad mastiff-bat	Costa Rica – Surinam, Trinidad; ref. 4.100

Cheiromeles

C. parvidens		C Sulawesi, Philippines
C. torquatus	Hairless bat	Malaya – Java, Borneo, Philippines

ORDER PRIMATES

Primates; *c.* 181 species; S, C America, Africa, SE Asia; forest (savanna); mainly arboreal; *.

Family Cheirogaleidae

Mouse-lemurs, dwarf lemurs; *c.* 7 species; Madagascar; forest; ref. 5.4; *.

Microcebus; mouse-lemurs.

M. murinus	Lesser mouse-lemur	S, W Madagascar
M. rufus	Russet mouse-lemur (Brown mouse-lemur)	E Madagascar

Mirza; (*Microcebus*).

M. coquereli	Coquerel's dwarf lemur	W Madagascar

Cheirogaleus; dwarf lemurs.

C. major	Greater dwarf lemur	N, E Madagascar
C. medius	Fat-tailed dwarf lemur	S, W Madagascar

Allocebus

A. trichotis	Hairy-eared dwarf lemur	NE Madagascar

Phaner
P. *furcifer*   Fork-marked lemur   N, W Madagascar

Family Lemuridae

Large lemurs; *c.* 10 species; Madagascar; mainly forest; ref. 5.4; *.

Lemur
L. *catta*   Ring-tailed lemur   S Madagascar
L. *coronatus*   Crowned lemur   N Madagascar
L. *fulvus*   Brown lemur   Madagascar
   (*albifrons*)
L. *macaco*   Black lemur   NW Madagascar
L. *mongoz*   Mongoose-lemur   NW Madagascar,
                                                   Comoro Is
L. *rubriventer*   Red-bellied lemur   E Madagascar

Hapalemur; gentle lemurs.
H. *griseus*   Grey gentle lemur   E, N, WC Madagascar
H. *simus*   Broad-nosed gentle lemur   EC Madagascar

Varecia; (*Lemur*).
V. *variegata*   Ruffed lemur   E Madagascar

Lepilemur
L. *mustelinus*   Weasel-lemur, Sportive   Madagascar
   (*ruficaudatus*)   lemur
   (*septentrionalis*)
   (*dorsalis*)
   (*edwardsi*)
   (*leucopus*)
   (*microdon*)

Family Indriidae

Leaping lemurs; 4 species; Madagascar; forest; arboreal; ref. 5.4; *.

Avahi; (*Lichanotus*).
A. *laniger*   Woolly lemur   E, NW Madagascar

Propithecus; sifakas.
P. *diadema*   Diadem sifaka   E, N Madagascar
P. *verreauxi*   Verreaux's sifaka   W, S Madagascar

Indri
I. *indri*   Indri   NE Madagascar

Family Daubentoniidae

One species; *.

Daubentonia
D. madagascariensis   Aye-aye   N Madagascar; forest

Family Lorisidae

Lorises, bushbabies; *c.* 14 species; SE Asia, Africa; forest, (savanna); arboreal, omnivorous; *.

Loris
L. tardigradus   Slender loris   S India, Sri Lanka

Nycticebus; slow lorises.
N. coucang   Slow loris   Assam – Java, Borneo,
                                                     Philippines
N. pygmaeus   Pygmy slow loris   Indochina

Perodicticus
P. potto   Potto   Guinea – Zaire – Kenya;
                                                     forest

Arctocebus
A. calabarensis   Angwantibo   R Niger – R Zaire; forest

Galago
G. alleni   Allen's bushbaby   R Niger – R Zaire; forest
G. granti   Grant's bushbaby   Mozambique, etc.
G. senegalensis   Lesser bushbaby   Senegal – Somalia –
                                                     Transvaal; savanna
G. zanzibaricus   Zanzibar bushbaby   Kenya – Mozambique,
                                                     Zanzibar

Otolemur; (Galago); greater bushbabies.
O. crassicaudatus   Angola – Kenya – Natal
O. garnettii   S Somalia –
                                                     N Mozambique

Euoticus; (Galago); needle-clawed bushbabies; C Africa; forest.
E. elegantulus   Western needle-clawed   R Niger – R Zaire
                         bushbaby
E. inustus   Eastern needle-clawed   E Zaire
                         bushbaby

Galagoides; (*Galago*).

G. *demidoff* (*demidovii*)	Demidoff's galago	Senegal – Tanzania; forest

Family Tarsiidae

Tarsiers; 3 species; Sumatra, Borneo, Sulawesi; Philippines, etc.; *.

Tarsius

T. *bancanus*	Western tarsier (Horsfield's tarsier)	Sumatra, Borneo
T. *spectrum*	Spectral tarsier	Sulawesi
T. *syrichta*	Philippine tarsier	Mindanao, Philippines

Family Callithricidae

Marmosets, tamarins; *c.* 18 species; tropical S, C America; forest, (savanna); arboreal; ref. 5.3; *.

Callithrix; (*Hapale*); marmosets.

C. *argentata*	Silvery marmoset (Black-tailed marmoset)	Brazil, Bolivia, S of Amazon
C. *humeralifer* (*chrysoleuca*) (*santaremensis*)	Santarem marmoset	R Madiera – R Tapajos, Amazon Basin
C. *jacchus* (*flaviceps*) (*penicillata*)	Common marmoset (White-eared marmoset) (White-headed marmoset) (Black-plumed marmoset)	E Brazil

Cebuella

C. *pygmaea*	Pygmy marmoset	Upper Amazon Basin

Saguinus; (*Hapale, Leontocebus*); tamarins.

S. *bicolor* (*martinsi*)	Bare-faced tamarin (Martin's tamarin)	NC Amazon Basin
S. *fuscicollis*	Saddle-back tamarin	Upper Amazon Basin
S. *imperator*	Emperor tamarin	W Brazil, E Peru, N Bolivia
S. *inustus*	Mottle-faced tamarin	NW Brazil, E Colombia
S. *labiatus*	White-lipped tamarin	C Amazon
S. *leucopus*	White-footed tamarin	N Colombia
S. *midas* (*tamarin*)	Red-handed tamarin (Negro tamarin)	N Brazil, Guianas
S. *mystax*	Moustached tamarin	N Peru, NW Brazil (S of Amazon)

S. nigricollis (*graellsi*)	Black and red tamarin	NW Brazil, etc.
S. oedipus (*geoffroyi*)	Cotton-top tamarin, Geoffroy's tamarin	Colombia, Panama

Leontopithecus; (*Leontocebus, Leontideus*); golden tamarins; ref. 5.5.

L. chrysomelas	Golden-headed tamarin	SE Bahia, Brazil
L. chrysopygus	Golden-rumped tamarin	Sao Paulo, Brazil
L. rosalia	Golden lion tamarin	SE Brazil; coastal forest

Callimico

C. goeldii	Goeldi's marmoset	Upper Amazon Basin

Family Cebidae

New-world monkeys; *c.* 35 species; S, C America; forest, arboreal; ref. 5.1; *.

Cebus; capuchin monkeys; Honduras – N Argentina.

C. albifrons	Brown pale-fronted capuchin	Upper Amazon, Venezuela, [Trinidad]
C. apella	Black-capped capuchin (Brown capuchin)	Colombia, Venezuela – N Argentina
C. capucinus	White-throated capuchin	Honduras – Colombia
C. olivaceus (*nigrivittatus*)	Weeper capuchin	Venezuela – mouth of Amazon

Aotus; night monkeys, douroucoulis; as many as 9 species have recently been recognized but their status as species or subspecies is uncertain; ref. 5.6.

A. azarae (*infulatus*) (*miconax*) (*nancymai*) (*nigriceps*)	Southern night monkey	Amazon – Paraguay
A. trivirgatus (*brumbacki*) (*lemurinus*) (*vociferans*)	Northern night monkey	Panama – Amazon

Callicebus; titis.

C. moloch	Dusky titi	Colombia – Bolivia
C. personatus	Masked titi	SE Brazil
C. torquatus	Widow monkey (White-handed titi)	Amazon – Colombia, Venezuela

Saimiri; squirrel monkeys; ref. 5.7.

S. boliviensis		Upper Amazon
S. oerstedii	Red-backed squirrel-monkey	Costa Rica, Panama
S. sciureus	Common squirrel-monkey	Colombia – Amazon Basin
S. ustus		C Brazil, S of Amazon

Pithecia; sakis.

P. albicans		Central Amazon
P. hirsuta		Amazon Basin
P. monachus	Monk saki	Upper Amazon Basin
P. pithecia	White-faced saki	Orinoco – Lower Amazon

Cacajao; uakaris; Amazon Basin.

C. calvus	White uakari (Bald uakari)	C Amazon
C. melanocephalus	Black-headed uakari	C Amazon – S Venezuela
C. rubicundus (*calvus*)	Red uakari	Upper Amazon Basin

Chiropotes; bearded sakis.

C. albinasus	White-nosed saki	S Amazon Basin
C. satanas	Black-bearded saki (Black saki)	Amazon – Guianas, Venezuela

Alouatta; howler monkeys; S Mexico – N Argentina.

A. belzebul	Black and red howler	Lower Amazon
A. caraya	Black howler	N Argentina – S Brazil
A. fusca	Brown howler	SE Brazil, Bolivia
A. palliata	Mantled howler	S Mexico – Ecuador
A. seniculus	Red howler	Colombia – mouth of Amazon – Bolivia
A. villosa (*pigra*)	Guatemalan howler	S Mexico – Guatemala

Ateles; spider monkeys; C Mexico – Bolivia.

A. belzebuth	Long-haired spider monkey	Colombia, NW Brazil, etc.
A. fusciceps	Brown-headed spider monkey	Panama – Ecuador
A. geoffroyi	Black-handed spider monkey	C Mexico – Colombia
A. paniscus	Black spider monkey	NE Brazil, Guianas – Bolivia

Brachyteles

B. arachnoides	Woolly spider monkey	SE Brazil; coastal forest

Lagothrix; woolly monkeys.

L. flavicauda	Yellow-tailed woolly monkey (Hendee's woolly monkey)	NE Peru
L. lagothricha	Common woolly monkey	C, Upper Amazon Basin

Family Cercopithecidae

Old-world monkeys; *c.* 78 species; Africa, S, E Asia; forest, (savanna); ref. 5.1; *.

Subfamily Cercopithecinae

Macaca; (*Cynopithecus*); macaques; S, E Asia, (NW Africa); mainly forest; terrestrial, arboreal.

M. arctoides (*speciosa*)	Bear macaque (Stump-tailed macaque)	Burma, S China – Malaya
M. assamensis	Assam macaque	Nepal – Thailand
M. cyclopis	Taiwan macaque	Taiwan
M. fascicularis (*irus*)	Crab-eating macaque	S Burma – Java, Borneo, Philippines
M. fuscata (*speciosa*)	Japanese macaque	Japan (except Hokkaido)
M. maurus	Moor macaque	S Sulawesi
M. mulatta	Rhesus macaque	E Afghanistan – S, E China
M. nemestrina	Pigtail macaque	Assam – Sumatra, Borneo
M. nigra (*nigrescens*)	Celebes ape (Black ape)	NE Sulawesi
M. ochreata (*brunnescens*)	Booted macaque	SE Sulawesi
M. radiata	Bonnet macaque	S India
M. silenus	Liontail macaque	SW India
M. sinica	Toque macaque	Sri Lanka
M. sylvanus	Barbary ape	Morocco, N Algeria
M. thibetana	Père David's macaque (Tibetan stump-tailed macaque)	Sichuan, Fujian, China
M. tonkeana (*hecki*)	Tonkean macaque	C, N Sulawesi

Cercocebus; (*Lophocebus*); mangabeys; W, C, (E) Africa; forest.

C. albigena	White-cheeked mangabey	Cameroun – Kenya, Tanzania

C. aterrimus	Black mangabey (Crested mangabey)	Zaire, Angola
C. galeritus (*agilis*)	Agile mangabey (Tana River mangabey)	Zaire, etc., E Kenya, S Tanzania
C. torquatus (*atys*)	White-collared mangabey (Sooty mangabey)	Sierra Leone – Congo Rep.

Papio; savanna baboons; Africa except NW, (S Arabia); savanna.

P. anubis	Olive baboon	R Niger – Kenya
P. cynocephalus	Yellow baboon	Somalia – Mozambique – Angola
P. hamadryas	Hamadryas baboon	Ethiopia, Somalia, S Arabia
P. papio	Guinea baboon	Senegal, Guinea
P. ursinus	Chacma baboon	Zambia, Angola – S Africa

Mandrillus; (*Papio*); forest baboons; WC Africa; forest.

M. leucophaeus	Drill	SE Nigeria – Sanaga R (Cameroun)
M. sphinx	Mandrill	Sanaga R – Gabon

Theropithecus

T. gelada	Gelada	Ethiopia; montane

Cercopithecus; guenons; Africa S of Sahara; forest, (savanna).

C. aethiops (*pygerythrus*) (*tantalus*) (*sabaeus*)	Savanna monkey (Green monkey) (Vervet) (Grivet)	Senegal – Somalia – S Africa; savanna
C. ascanius	Schmidt's guenon, etc.	W Kenya – Zambia, Angola
C. campbelli	Campbell's monkey	Gambia – Ghana
C. cephus	Moustached monkey	Cameroun – Gabon, etc.
C. denti (*wolfi*)	Dent's monkey	NE, E Zaire
C. diana	Diana monkey	Sierra Leone – Ghana
C. dryas	Dryas monkey	C Zaire
C. erythrogaster	Red-bellied monkey	SW Nigeria
C. erythrotis	Red-eared monkey	Nigeria, Cameroun
C. hamlyni	Owl-faced monkey	E Zaire, etc.
C. lhoesti	L'Hoest's monkey	SW Uganda, NE Zaire, etc.
C. mitis (*albogularis*)	Diademed monkey (Sykes' monkey)	Somalia – Zaire – S Africa
C. mona	Mona monkey	Ghana – Cameroun

C. neglectus	De Brazza's monkey	Cameroun – Ethiopia – Angola
C. nictitans	Greater white-nosed monkey	Guinea – Zaire
C. petaurista	Lesser white-nosed monkey	Guinea – Togo
C. pogonias	Crowned guenon	SE Nigeria – N, W Zaire, etc.
C. preussi (*lhoesti*)	Preuss's monkey	W Cameroun, etc.
C. salongo	Zaire Diana monkey	C Zaire; ref. 5.8
C. wolfi	Wolf's monkey	C Zaire

Miopithecus; (*Cercopithecus*).

M. talapoin	Talapoin	Gabon – W Angola; forest

Allenopithecus; (*Cercopithecus*).

A. nigroviridis	Allen's swamp monkey (Blackish-green guenon)	Zaire, Congo Rep.; forest

Erythrocebus; (*Cercopithecus*).

E. patas	Patas monkey (Red monkey)	Senegal – Ethiopia – Tanzania; savanna, steppe

Subfamily Colobinae, see ref. 5.9 for alternative classification.

Colobus; (*Procolobus*); colobus monkeys; W, C, E Africa; forest.

C. angolensis	Angolan colobus	Angola – Kenya
C. badius (*rufomitratus*)	Red colobus (Bay colobus)	Gambia – Tanzania
C. guereza	Guereza (Eastern black-and-white colobus)	E Nigeria – Ethiopia – Tanzania
C. kirkii	Kirk's colobus	Zanzibar
C. polykomos	King colobus (Western black-and-white colobus)	Gambia – Togo
C. satanas	Black colobus	Cameroun – Gabon

Procolobus; (*Colobus*).

P. verus	Olive colobus	Sierra Leone – Nigeria; forest

Pygathrix; (*Rhinopithecus*).

P. avunculus	Tonkin snub-nosed monkey	N Vietnam
P. brelichi	Brelich's snub-nosed monkey	Guizhou, S China; close to *P. roxellana*
P. nemaeus (*nigripes*)	Douc langur	Indochina, Hainan
P. roxellana (*bieti*)	Chinese snub-nosed monkey (Golden monkey) (Snow monkey)	Yunnan, Sichuan, S China

Simias; (*Nasalis*).

S. concolor	Pig-tailed langur	Mentawai Is (Sumatra)

Nasalis

N. larvatus	Proboscis monkey	Borneo

Presbytis; (*Semnopithecus*); leaf monkeys, surelis; SE Asia; forest.

P. comata (*hosei, aygula*) (*thomasi*)	Sunda leaf monkey (Grizzled leaf monkey)	Sumatra, Java, Borneo
P. cristata	Silvered leaf monkey	Indochina – Java, Borneo
P. entellus	Hanuman langur	India, etc., Sri Lanka
P. francoisi	François' monkey	S China, Indochina
P. frontata	White-fronted leaf monkey	Borneo
P. geei	Golden leaf monkey	Assam, Bhutan
P. johnii	Nilgiri langur (John's leaf monkey)	SW India; (in *P. vetulus?*)
P. melalophos (*femoralis*)	Banded leaf monkey (Mitred leaf monkey)	Thailand – Sumatra, Borneo
P. obscura	Dusky leaf monkey (Spectacled leaf monkey)	S Thailand, Malaya
P. phayrei	Phayre's leaf monkey	Burma, Thailand, etc.
P. pileata	Capped leaf monkey	Assam, Burma
P. potenziani	Mentawai leaf monkey	Mentawai Is (Sumatra)
P. rubicunda	Maroon leaf monkey	Borneo
P. vetulus (*senex*)	Purple-faced leaf monkey	Sri Lanka

Family Hylobatidae

Gibbons; 6 species; SE Asia; forest; *.

Hylobates; (*Nomascus, Symphalangus*); gibbons; SE Asia.

H. concolor	Crested gibbon (Black gibbon)	Indochina, Hainan

H. hoolock	Hoolock gibbon	Assam, Burma, Yunnan
H. klossii	Kloss's gibbon (Dwarf siamang)	Mentawai Is (Sumatra)
H. lar (*agilis*) (*moloch*) (*muelleri*)	Common gibbon (Lar gibbon) (Agile gibbon) (Silvery gibbon)	S Burma – Sumatra, Java, Borneo
H. pileatus	Pileated gibbon	SE Thailand, etc.
H. syndactylus	Siamang	Malaya, Sumatra

Family Pongidae

Apes; 4 species; W, C Africa, SE Asia; forest; *.

Pongo

| *P. pygmaeus* | Orang-utan | Sumatra, Borneo |

Pan; chimpanzees.

| *P. paniscus* | Pygmy chimpanzee (Bonobo) | Zaire (S of Zaire R.) |
| *P. troglodytes* | Chimpanzee | Guinea – Zaire – Uganda, Tanzania |

Gorilla

| *G. gorilla* | Gorilla | SE Nigeria – W Zaire; E Zaire, etc. |

Family Hominidae

Homo

| *H. sapiens* | Man | Worldwide |

ORDER CARNIVORA

Carnivores; *c.* 266 species; worldwide; terrestrial, arboreal, aquatic carnivores and omnivores.

Family Canidae

Dogs, foxes; *c.* 35 species; Eurasia, Americas, Africa, [Australia].

Canis; dogs, jackals; Africa, Eurasia, N, C America; desert – open forest.

| *C. adustus* | Side-striped jackal | Senegal – Somalia – SW Africa; savanna, steppe |
| *C. aureus* | Golden jackal | Senegal – Thailand, Sri Lanka; steppe – open forest |

C. latrans	Coyote	Alaska, Canada, USA (except SE) – Costa Rica; desert – grassland
C. lupus	Wolf	Palaearctic (except N Africa), India, Alaska, Canada, Mexico, † USA; tundra, steppe, open forest; ancestor of domestic dogs, *C. familiaris*; *
C. mesomelas	Black-backed jackal	S, E Africa – Cameroun; savanna, steppe
C. rufus (*niger*)	Red wolf	SE Texas, etc.; *
C. simensis	Simian jackal (Simian fox)	Ethiopia; high montane grassland; *

Alopex; (*Vulpes*).

A. lagopus	Arctic fox	Eurasia, N America; tundra

Vulpes; (*Fennecus, Urocyon*); foxes; Africa, Eurasia, N America; desert – forest.

V. bengalensis	Bengal fox	India, Pakistan, Nepal; steppe – open forest
V. cana	Blanford's fox	Afghanistan etc.; montane steppe; *
V. chama	Cape fox	SW Africa; steppe
V. cinereoargenteus	Grey fox	S Canada – Venezuela
V. corsac	Corsac fox	S Russia – Manchuria; steppe
V. ferrilata	Tibetan fox	Tibet, etc.; steppe
V. littoralis	Island grey fox	Santa Barbara Is, California
V. macrotis	Kit fox	SW USA, N Mexico; desert, steppe; (in *V. velox*?)
V. pallida	Pale fox	Senegal – Somalia; desert
V. rueppellii	Sand fox	Morocco – Afghanistan; desert, steppe
V. velox	Swift fox	C USA; grassland; (*)
V. vulpes (*fulva*)	Red fox	Palaearctic – Indochina, Canada, USA, [Australia]; forest; includes domesticated forms, e.g. silver fox

V. zerda	Fennec fox	Morocco – Arabia; desert; (*)

Dusicyon; (*Atelocyon, Cerdocyon, Lycalopex*); S America; forest – desert; classification provisional.

D. australis†	Falkland Island wolf	Falkland Is (extinct)
D. culpaeus (*culpaeolus*)	Colpeo fox	Ecuador – Tierra del Fuego; *
D. griseus (*fulvipes*)	Argentine grey fox	Argentina, Chile; *
D. gymnocercus	Pampas fox	Argentina – Paraguay
D. microtis	Small-eared zorro	Amazon, Orinoco basins; forest; *
D. sechurae	Sechura fox	NW Peru, SW Ecuador; desert
D. thous	Common zorro (Crab-eating fox)	Colombia N Argentina; savanna, open forest
D. vetulus	Hoary fox	C, S Brazil

Nyctereutes

N. procyonoides	Raccoon-dog	Indochina – SE Sibera, Japan, [E, C Europe]; forest

Chrysocyon

C. brachyurus	Maned wolf	E Brazil – N Argentina; grassland; *

Speothos

S. venaticus	Bush dog	Panama – SE Brazil, E of Andes; forest; *

Cuon

C. alpinus	Dhole (Red dog)	Altai, Pamirs – E Siberia – Java; forest; *

Lycaon

L. pictus	Hunting dog	Ivory Coast – Somalia – S Africa; steppe, savanna; *

Otocyon

O. megalotis	Bat-eared fox	S Africa – S Zambia; Tanzania – Ethiopia; grassland, steppe

Family Ursidae

Bears; 7 species; Eurasia, N, (S) America; mainly forest; omnivores, (carnivores); *.

Tremarctos
T. ornatus	Spectacled bear	Venezuela – Bolivia; forest

Selenarctos; (*Ursus*).
S. thibetanus	Asiatic black bear	Afghanistan – Indochina – E Siberia, Japan; forest

Ursus
U. americanus	American black bear	C Mexico – Alaska; forest
U. arctos	Brown bear	Eurasia N of Himalayas,
(*horribilis*)	(Grizzly bear)	N America; mainly forest

Thalarctos; (*Ursus*).
T. maritimus	Polar bear	Arctic Ocean

Helarctos
H. malayanus	Sun bear	Burma – Sumatra, Borneo; forest

Melursus
M. ursinus	Sloth bear	India, Sri Lanka; forest

Family Procyonidae

Raccoons etc.; *c.* 13 species; S, C, (N) America; forest; omnivores.

Bassariscus; (*Jentinkia*).
B. astutus	Ring-tailed cat (Cacomistle)	Oregon – S Mexico
B. sumichrasti	Central American cacomistle	S Mexico – Panama; (*)

Procyon; raccoons; insular forms in the W Indies and Mexico are sometimes treated as distinct species but are likely to represent *P. lotor.*
P. cancrivorus	Crab-eating raccoon	Costa Rica – N Argentina
P. lotor	Common raccoon	S Canada – Panama,
(*gloveralleni*)		[Bahamas, Guadeloupe]
(*insularis*)		
(*minor*)		
(*pygmaeus*)		

Nasua; coatis, coatimundis; S, C, (N) America.

N. nasua (*narica*)	Coati	S USA – S America except Patagonia; (*)
N. nelsoni	Cozumel coati	Cozumel I, Yucatan, Mexico

Nasuella

N. olivacea	Little coatimundi	Venezuela – Ecuador

Potos

P. flavus	Kinkajou	E Mexico – S Brazil; forest

Bassaricyon; olingos; C, northern S America; possibly all one species.

B. alleni		Ecuador, Peru
B. beddardi		Guyana, etc.
B. gabbii	Bushy-tailed olingo	Nicaragua – Ecuador, Venezuela; (*)
B. lasius	Harris' olingo	Costa Rica
B. pauli	Chiriqui olingo	W Panama

Family Ailuropodidae

Pandas; 2 species; Himalayas, S China; montane forest; herbivores; *Ailuropoda* is often included in the Ursidae and *Ailurus* (with less reason) in the Procyonidae; *.

Ailurus

A. fulgens	Lesser panda (Red panda)	Nepal – W Burma – Sichuan

Ailuropoda

A. melanoleuca	Giant panda	Sichuan etc., S China; bamboo forest

Family Mustelidae

Weasels, etc., *c.* 63 species; Americas, Eurasia, Africa; forest – desert, (aquatic); carnivores, (omnivores).

Mustela; (*Grammogale, Lutreola, Putorius*); weasels; Americas, Eurasia, (N Africa); forest, steppe, tundra.

M. africana	Amazon weasel	Amazon Basin (sic)
M. altaica	Mountain weasel	Himalayas – Altai – Korea
M. erminea	Stoat (Ermine)	Europe – E Siberia, Japan, Alaska – N Greenland – N USA; [New Zealand]; forest, tundra

M. eversmanni	Steppe polecat	E Europe – Manchuria – Tibet; grassland, steppe
M. felipei		Colombia; montane
M. frenata	Long-tailed weasel	S Canada – Venezuela – Peru
M. kathiah	Yellow-bellied weasel	W Himalayas – S China
M. lutreola	European mink	W Siberia, E Europe, (W Europe); riversides, marshes
M. lutreolina		Java
M. nigripes	Black-footed ferret	Alberta – N Texas; grassland; *
M. nivalis (*rixosa*) (*minuta*)	Weasel (Lesser weasel)	N Africa, W Europe – E Siberia; Japan, Alaska – NE USA, [New Zealand]
M. nudipes	Malaysian weasel	Malaya, Sumatra, Borneo
M. putorius	Polecat	Europe; forest; probable ancestor of domestic ferret, *M. furo*
M. sibirica	Siberian weasel (Kolinsky)	Siberia – Himalayas, Thailand; Japan
M. strigidorsa	Back-striped weasel	Nepal – Thailand
M. vison	American mink	Canada, USA, [Iceland, N, C Europe, Siberia]; rivers, lakes; includes domesticated mink

Vormela

V. peregusna	Marbled polecat	SE Europe – W China; steppe, grassland

Martes; (*Charronia*); martens; Eurasia, N America; forest; arboreal carnivores.

M. americana	American marten	Canada, N USA
M. flavigula (*gwatkinsi*)	Yellow-throated marten	E Siberia – Java, Borneo; S India
M. foina	Beech marten	W Europe – Himalayas, Altai
M. martes	Pine marten	W Europe – W Siberia
M. melampus	Japanese marten	Japan (except Hokkaido), S Korea
M. pennanti	Fisher	Canada, N USA
M. zibellina	Sable	Siberia, Hokkaido

Eira; (*Galera*, *Tayra*).

| *E. barbara* | Tayra | NE Mexico – Argentina |

Galictis; (*Grison*, *Grisonella*); grisons; S Mexico – Brazil.

| *G. cuja* | Little grison | Peru – Uruguay – Chile |
| *G. vittata* | Greater grison | S Mexico – Peru – E Brazil |

Lyncodon

| *L. patagonicus* | Patagonian weasel | Argentina, Chile; grassland |

Ictonyx; (*Zorilla*).

| *I. striatus* | Zorilla
(Striped polecat) | Senegal – Ethiopia –
S Africa; steppe,
savanna |

Poecilictis

| *P. libyca* | Saharan striped weasel | Sahara, etc.; desert, steppe |

Poecilogale

| *P. albinucha* | White-naped weasel | S Africa – Zaire, Uganda;
savanna |

Gulo

| *G. gulo*
(*luscus*) | Wolverine
(Glutton) | Scandinavia, Siberia,
Alaska, Canada,
W USA; con. forest,
tundra |

Mellivora

| *M. capensis* | Ratel
(Honey badger) | N India – Arabia, Africa S
of Sahara; steppe,
savanna; (*) |

Meles

| *M. meles* | Eurasian badger | Europe – Japan – S China;
forest, grassland |

Arctonyx

| *A. collaris* | Hog-badger | N China – NE India –
Sumatra |

Mydaus; (*Suillotaxus*); stink badgers.

| *M. javanensis* | Sunda stink badger | Sumatra, Java, Borneo |
| *M. marchei* | Palawan stink badger | Palawan, Calamian Is,
Philippines |

Taxidea
T. taxus   American badger   SW Canada – C Mexico;
                                                         grassland, steppe

Melogale; (*Helictis*); ferret-badgers; SE Asia, grassland, open forest.
M. everetti   Everett's ferret-badger   Borneo
M. moschata   Chinese ferret-badger   NE India – S China –
                                                         Indochina; Java, Borneo
M. personata   Burmese ferret-badger   Nepal – Indochina; Java
  (*orientalis*)

Mephitis
M. macroura   Hooded skunk   SW USA – Costa Rica
M. mephitis   Striped skunk   S Canada – N Mexico

Spilogale; spotted skunks; N, C America.
S. putorius   Spotted skunk   SE, C USA – Costa Rica
  (*angustifrons*)
  (*gracilis*)
S. pygmaea   Pygmy spotted skunk   W, SW Mexico

Conepatus; hog-nosed skunks; C, S America.
C. chinga   Chile – Peru, S Brazil
  (*rex*)
C. humboldti   Patagonia – Paraguay; *
  (*castaneus*)
C. leuconotus   Eastern hog-nosed skunk   E Texas, E Mexico
C. mesoleucus   Hog-nosed skunk   S USA – Nicaragua
C. semistriatus   Striped hog-nosed skunk   S Mexico – Peru – E Brazil

Lutra; (*Lontra, Lutrogale*); river otters; Eurasia, Americas, Africa; rivers, lakes,
(sea-coast); *.
L. canadensis   Canadian otter   Canada, USA
L. felina   N Peru – S Chile; coastal
L. longicaudis   C Mexico – Uruguay
  (*annectens*)
L. lutra   European otter   Eurasia, N Africa, Sri
                                                         Lanka, Taiwan,
                                                         Sumatra, Java
L. maculicollis   Spotted-necked otter   Liberia – Ethiopia –
                                                         S Africa
L. perspicillata   Smooth-coated otter   Iraq; India – Sumatra,
                                                         Borneo
L. provocax   Chile, S Argentina
L. sumatrana   Hairy-nosed otter   Indochina – Java, Borneo

Pteronura
P. brasiliensis   Giant otter   Venezuela – Argentina;
                                                        rivers; *

Aonyx; (Amblonyx, Paraonyx); clawless otters; Africa, SE Asia; *.
A. capensis   African clawless otter   Senegal – Ethiopia –
                                                        S Africa
A. cinerea   Oriental small-clawed otter   India, S China – Java,
                                                        Borneo, Palawan
A. congica   Zaire clawless otter   Congo basin, etc., forest
   (microdon)

Enhydra
E. lutris   Sea otter   Bering Sea – California;
                                                        rocky coasts; *

Family Viverridae

Civets, etc.; c. 35 species; S Asia, Africa, Madagascar; forest – steppe; mainly
omnivores.

Poiana
P. richardsonii   African linsang   Sierra Leone – N Zaire;
                                                        forest

Genetta; genets; Africa, (Arabia, SW Europe); forest, savanna; arboreal;
classification very provisional; ref. 6.1.
G. abyssinica   Abyssinian genet   Ethiopia – Somalia
G. angolensis   Angolan genet   Angola – Mozambique
   (mossambicus)
G. cristata   Cameroun – SE Nigeria
   (servalina)
G. genetta   Small-spotted genet   Africa, SW Europe, Arabia
   (felina)   – Asia Minor; savanna
G. johnstoni   Johnston's genet   Liberia
G. pardina   Pardine genet   Gambia – Cameroun;
                                                        forest
G. rubiginosa   Rusty-spotted genet   Senegal – Somalia –
   (maculata)   Transvaal; savanna
G. servalina   Servaline genet   Cameroun – Kenya; forest
G. thierryi   Hausa genet   Senegal – Chad; guinea
                                                        savanna
G. tigrina   Large-spotted genet   S Africa
G. victoriae   Giant genet   Uganda, NE Zaire

Viverricula

V. indica (*malaccensis*)	Small Indian civet	India – S China – Java; Sri Lanka; [Madagascar, Zanzibar]

Osbornictis

O. piscivora	Congo water civet (Aquatic genet)	NE Zaire; lowland forest

Viverra; (*Civettictis*); civets; Africa, SE Asia.

V. civetta	African civet	Senegal – Somalia – Transvaal; (*)
V. megaspila	Large-spotted civet	India – Indochina – Malaya; (*)
V. tangalunga	Malay civet	Malaya, Sumatra, Borneo, Philippines
V. zibetha	Large indian civet	India – S China – Malaya

Prionodon; (*Pardictis*); linsangs; SE Asia; *.

P. linsang	Banded linsang	Thailand – Java, Borneo
P. pardicolor	Spotted linsang	Nepal – Indochina

Nandinia

N. binotata	African palm civet (Tree civet)	Guinea – S Sudan – Mozambique; forest

Arctogalidea

A. trivirgata	Three-striped palm civet	Assam – Java, Borneo

Paradoxurus

P. hermaphroditus (*philippinensis*)	Common palm civet (Toddy cat)	India – S China – Java, Borneo, Timor, Sulawesi, Philippines, Sri Lanka
P. jerdoni	Jerdon's palm civet	S India
P. zeylonensis	Golden palm civet	Sri Lanka

Paguma

P. larvata	Masked palm civet	Himalayas – China – Sumatra, Borneo, Taiwan

Macrogalidea

M. musschenbroekii	Brown palm civet	NE Sulawesi; *

Arctictis
A. binturong (whitei)	Binturong	Burma – Java, Borneo, Palawan

Fossa
F. fossa (fossana)	Malagasy civet	NW Madagascar; forest; *

Hemigalus; (Diplogale).
H. derbyanus	Banded palm civet	Malaya, Sumatra, Borneo; *
H. hosei	Hose's civet	Borneo

Chrotogale
C. owstoni	Owston's palm civet	Indochina

Cynogale
C. bennettii	Otter-civet	Indochina – Sumatra, Borneo; *

Eupleres
E. goudotii (major)	Falanouc	N Madagascar; forest; *

Cryptoprocta; family allocation dubious.
C. ferox	Fossa	Madagascar; forest, savanna; *

Family Herpestidae

Mongooses; c. 37 species; Africa, S Asia; forest – steppe; carnivores and omnivores; sometimes included in the Viverridae.

Galidea
G. elegans	Ring-tailed mongoose	Madagascar; forest

Galidictis
G. fasciata (striata)	Broad-striped mongoose	E Madagascar; forest

Mungotictis
M. decemlineata (lineata) (substriatus)	Narrow-striped mongoose	W, SW Madagascar; savanna

Salanoia

S. concolor (*olivacea*)	Salano	NE Madagascar; forest

Suricata

S. suricatta	Meerkat (Suricate)	S Africa – S Angola

Herpestes; (*Xenogale*, *Galerella*); common mongooses; Africa, S Asia, (SW Europe); forest – steppe; terrestrial.

H. auropunctatus (*javanicus*) (*palustris*)	Small Indian mongoose	E Arabia – Malaya; [Hawaii, W Indies, Guyana, etc.]
H. brachyurus	Short-tailed mongoose	Malaya, Sumatra, Borneo, Palawan
H. edwardsii	Indian grey mongoose	Arabia – Assam, Sri Lanka; [Malaya, Ryukyu Is, Mauritius]
H. fuscus	Indian brown mongoose	S India, Sri Lanka; (in *H. brachyurus*?)
H. hosei	Hose's mongoose	Borneo; (in *H. brachyurus*?)
H. ichneumon	Egyptian mongoose (Large grey mongoose)	Iberia; Arabia, etc.; Africa S of Sahara; savanna
H. javanicus	Javan mongoose	Indochina, Malaya, Java
H. naso	Long-nosed mongoose	Zaire – SE Nigeria; lowland forest
H. pulverentulus	Cape grey mongoose	S Africa – S Angola; steppe
H. sanguineus	Slender mongoose	Africa S of Sahara; savanna, steppe
H. semitorquatus	Collared mongoose	Sumatra, Borneo
H. smithii	Ruddy mongoose	India, Sri Lanka
H. urva	Crab-eating mongoose	Nepal – S China – Thailand
H. vitticollis	Stripe-necked mongoose	India, Sri Lanka

Helogale; dwarf mongooses; E, S Africa; savanna.

H. hirtula		Ethiopia, Somalia, Kenya
H. parvula		Ethiopia – Natal – Namibia

Dologale

D. dybowskii	Pousargues' mongoose	Cent. Afr. Rep. – W Uganda; savanna

Atilax
A. paludinosus   Marsh mongoose   Africa S of Sahara; rivers,
                    (Water mongoose)   marshes

Mungos
M. gambianus   Gambian mongoose   Gambia – Nigeria;
                                     savanna
M. mungo   Banded mongoose   Africa S of Sahara

Crossarchus; W, C Africa; forest; ref. 6.2.
C. alexandri   N Zaire, Uganda
C. ansorgei   C Zaire, N Angola
C. obscurus   Cusimanse   Sierra Leone – Ghana
                    (Long-nosed mongoose)
C. platycephalus   Benin – Cameroun

Liberiictis
L. kuhni   Liberian mongoose   Liberia

Ichneumia
I. albicauda   White-tailed mongoose   Africa S of Sahara,
                                      S Arabia; savanna,
                                      open forest

Bdeogale; (Galeriscus).
B. crassicauda   Bushy-tailed mongoose   Kenya – Mozambique
B. jacksoni   Jackson's mongoose   Kenya, Uganda
B. nigripes   Black-legged mongoose   SE Nigeria – Zaire –
                                      N Angola

Rhynchogale
R. melleri   Meller's mongoose   S Zaire – Transvaal

Cynictis
C. penicillata   Yellow mongoose   S Africa – S Angola

Paracynictis
P. selousi   Grey meerkat   Mozambique – Angola
                    (Selous' mongoose)

Family Hyaenidae

Hyaenas; 4 species; Africa, SW Asia; savanna – desert; terrestrial carnivores and scavengers.

Proteles; sometimes placed in separate family Protelidae.

P. cristatus	Aardwolf	S, E Africa; steppe; (*)

Crocuta

C. crocuta	Spotted hyaena	Africa S of Sahara; steppe, savanna

Hyaena

H. brunnea	Brown hyaena	S Africa – Zimbabwe; *
H. hyaena	Striped hyaena	Senegal – Tanzania – Turkestan – India; steppe, desert; (*)

Family Felidae

Cats; c. 36 species; Americas, Eurasia, Africa; forest – desert; carnivores; generic classification very unstable; *

Felis; (*Badiofelis, Herpailurus, Leptailurus, Mayailurus, Otocolobus, Prionailurus, Profelis, Puma*); small cats; Americas, Eurasia, Africa; mainly forest.

F. aurata	African golden cat	Senegal – Zaire – Kenya; forest
F. badia	Bay cat (Bornean red cat)	Borneo
F. bengalensis	Leopard cat	E Siberia – Baluchistan – Java, Borneo, Taiwan
F. bieti	Chinese desert cat	W China, S Mongolia; steppe
F. caracal	Caracal	Turkestan, NW India – Arabia; Africa; steppe, savanna
F. chaus	Jungle cat	Egypt – Indochina, Sri Lanka; dry forest
F. colocolo	Pampas cat	Ecuador – Patagonia; grassland, forest
F. concolor	Puma (Cougar)	S Canada – Patagonia; forest – steppe
F. geoffroyi	Geoffroy's cat	Borneo – Patagonia
F. guigna	Kodkod	C, S Chile, W Argentina; forest
F. iriomotensis	Iriomote cat	Iriomote I, Ryukyu Is; close to *F. bengalensis*
F. jacobita	Andean cat	S Peru – N Chile; montane steppe

F. manul	Pallas's cat	Iran – W China; montane steppe
F. margarita	Sand cat	Sahara – Baluchistan; desert
F. marmorata	Marbled cat	Nepal – Malaya, Sumatra, Borneo
F. nigripes	Black-footed cat (Small spotted cat)	S Africa, Botswana, Namibia; steppe, savanna
F. pardalis	Ocelot	Arizona N Argentina; forest, scrub
F. planiceps	Flat-headed cat	Malaya, Sumatra, Borneo
F. rubiginosa	Rusty-spotted cat	S India, Sri Lanka
F. serval	Serval	Africa; savanna
F. silvestris (*catus*) (*lybica*)	Wild cat	W Europe – India; Africa; open forest, savanna, steppe; ancestor of domestic cat *F. catus*
F. temminckii	Asiatic golden cat	Nepal – S China – Sumatra; forest
F. tigrina	Littled spotted cat (Oncilla)	Costa Rica – N Argentina; forest
F. viverrina	Fishing cat	India – S China – Java; forest, swamps
F. wiedii	Margay	N Mexico – N Argentina; forest
F. yagouaroundi	Jaguarundi	Arizona – C Argentina; forest, scrub

Lynx; (*Felis*); lynxes; N America, N Eurasia; *.

L. canadensis (*lynx*)	Canadian lynx	Canada, Alaska, N USA; con. forest
L. lynx (*pardina*)	Eurasian lynx	W Europe – Siberia; con. forest
L. rufus	Bobcat	S Canada – S Mexico

Panthera; (*Leo, Tigris, Uncia*); big cats; Americas, Asia, Africa; mainly forest.

P. leo	Lion	Africa S of Sahara; NW India; († NW Africa, SW Asia); savanna, steppe
P. onca	Jaguar	N Mexico – N Argentina; († S USA, Uruguay, C Argentina); forest

P. pardus	Leopard (Panther)	SE Siberia – Java – Asia Minor; Africa; forest
P. tigris	Tiger	SE Siberia – Java – Caucasus; forest
P. uncia	Snow leopard (Ounce)	Altai – Himalayas; montane steppe

Neofelis; (*Panthera*).

N. nebulosa	Clouded leopard	Nepal – S China – Sumatra, Borneo, Taiwan; forest

Acinonyx

A. jubatus	Cheetah	Baluchistan – Arabia; Africa; steppe, savanna

Family Otariidae

Eared seals, sealions; *c.* 14 species; S Hemisphere, N Pacific; formerly in order Pinnipedia but now considered related to the Ursidae; refs. 6.3, 4.

Arctocephalus; (*Arctophoca*); southern fur seals; S Hemisphere – California; *.

A. australis	South American fur seal	Brazil – Peru
A. forsteri	New Zealand fur seal	S New Zealand, etc., S Australia
A. galapagoensis	Galapagos fur seal	Galapagos Is
A. gazella	Antarctic fur seal	Islands S of Antarctic Convergence
A. philippii	Juan Fernandez fur seal	Juan Fernandez Is
A. pusillus (*doriferus*) (*tasmanicus*)	Afro-Australian fur seal	S Africa, S Australia, Tasmania
A. townsendi	Guadalupe fur seal	S California – W Mexico
A. tropicalis	Subantarctic fur seal (Amsterdam Island fur seal)	Islands N of Antarctic Convergence

Callorhinus

C. ursinus	Northern fur seal	Bering, Okhotsk Seas

Zalophus

Z. californianus	Californian sealion	California, Galapagos, († Japan)

Eumetopias

E. jubatus	Northern sealion (Steller's sealion)	N Japan – California

Otaria
O. byronia   South American sealion   S Brazil – Peru
  (*flavescens*)   (Southern sealion)

Neophoca
N. cinerea   Australian sealion   SW, S Australia

Phocarctos; (*Neophoca*).
P. hookeri   New Zealand sealion   S of New Zealand
                      (Hooker's sealion)

Family Odobenidae

One species.

Odobenus
O. rosmarus   Walrus   Arctic ocean; (*)

Family Phocidae

Earless seals; 19 species; polar, temperate, some tropical seas; formerly in order Pinnipedia but probably more closely related to the Mustelidae than to the Otariidae; refs. 6.3, 4.

Phoca; (*Pusa, Histriophoca, Pagophilus*); N temperate and polar seas.
P. caspica   Caspian seal   Caspian Sea
P. fasciata   Ribbon seal   N Pacific; subarctic
                      (Banded seal)
P. groenlandicus   Harp seal   N Atlantic; subarctic;
                      (Greenland seal)   breed on ice
P. hispida   Ringed seal   Arctic Ocean, N Pacific,
                                                      Baltic; breed on ice; (*)
P. largha   Larga seal   NW Pacific; breed on ice
P. sibirica   Baikal seal   Lake Baikal
P. vitulina   Common seal   N Atlantic, N Pacific;
  (*kurilensis*)   (Harbour seal)   temperate, subarctic
  (*richardii*)   (Spotted seal)

Halichoerus
H. grypus   Grey seal   Newfoundland, etc.,
                      (Atlantic seal)   Britain – White Sea,
                                                      Baltic

Erignathus
E. barbatus   Bearded seal   Arctic Ocean; breed on ice

Lobodon
L. carcinophagus   Crabeater seal   Antarctic; edge of pack ice

Ommatophoca
O. rossi   Ross seal   Antarctic; edge of pack ice

Hydrurga
H. leptonyx   Leopard-seal   Southern Ocean

Leptonychotes
L. weddelli   Weddell seal   Antarctic; edge of ice

Monachus; monk seals; northern tropical and warm temperate zones; *.
M. monachus   Mediterranean monk seal   Canary Is –
                                                 Mediterranean, Black
                                                 Sea
M. schauinslandi   Hawaiian monk seal   Hawaii
M. tropicalis †   Caribbean monk seal   Formerly Caribbean:
                                                 probably extinct

Mirounga; elephant-seals; *.
M. angustirostris   Northern elephant-seal W coast of N America
M. leonina   Southern elephant-seal Subantarctic Islands,
                                                 Patagonia

Cystophora
C. cristata   Hooded seal   N Atlantic; edge of ice
                      (Bladder-nose)

ORDER CETACEA

Whales, dolphins, porpoises; c. 78 species; all oceans, some tropical rivers; ref. 7.1, 7.2; *.

SUBORDER ODONTOCETI

Toothed whales; c. 66 species; all oceans, some tropical rivers.

Family Platanistidae

River dolphins; c. 5 species; S America, S, E Asia; rivers, coastal waters; fish-eaters; *.

Pontoporia
P. blainvillei   La Plata dolphin   S Brazil – N Argentina;
                      (Franciscana)   coasts, estuaries

Inia
I. geoffrensis — Boutu — Amazon, upper Orinoco
   (Amazon porpoise)

Lipotes
L. vexillifer — White fin dolphin — Yangtzekiang R, Tungting L, China

Platanista; (*Susu*).
P. gangetica — Ganges dolphin — Ganges, Brahmaputra
P. minor — Indus dolphin — Indus & tributaries
  (*indi*)

Family Delphinidae

Marine dolphins; *c.* 34 species; all oceans, some tropical rivers; mainly fish-eaters; *.

Steno
S. bredanensis — Rough-toothed dolphin — Tropical, warm temperate seas

Sotalia
S. fluviatilis — Tucuxi — Panama – S Brazil; Amazon; coasts, rivers
  (*brasiliensis*)
  (*guianensis*)

Sousa; hump-backed dolphins; tropical coasts, estuaries, large rivers.
S. chinensis — Indo-Pacific hump-backed dolphin — Indian Ocean, SW Pacific, lower Yangtzekiang
  (*borneensis*)
  (*lentiginosus*)
  (*plumbea*)
S. teuszii — Atlantic hump-backed dolphin — Mauritania – Angola

Stenella; tropical and warm temperate seas; classification provisional.
S. attenuata — Bridled dolphin — Tropical oceans
  (*dubia*) — (Spotted dolphin)
  (*frontalis*)
  (*graffmani*)
S. clymene — Atlantic spinner dolphin — Atlantic
S. coeruleoalba — Striped dolphin — Atlantic, Pacific; tropical, temperate
  (*styx*) — (Euphrosyne dolphin)
S. longirostris — Spinner dolphin — Tropical, warm temperate seas
  (Long-beaked dolphin)

S. plagiodon   Atlantic spotted dolphin   Atlantic; tropical, warm
                                                          temperate

Delphinus
D. delphis   Common dolphin   Temperate, tropical seas;
  (*bairdii*)   Black Sea
D. tropicalis   Tropical dolphin   Indian Ocean, etc., tropical

Tursiops
T. truncatus   Bottle-nosed dolphin   Atlantic, Indian Oc.,
  (*gilli*)   Pacific; tropical,
                                                          temperate

Lissodelphis; right whale dolphins; Pacific, S Atlantic.
L. borealis   Northern right whale   N Pacific; temperate
                          dolphin
L. peronii   Southern right whale   S Atlantic, S Pacific;
                          dolphin   temperate, subtropical

Lagenodelphis
L. hosei   Fraser's dolphin   Pacific, Indian Oc.,
                                                          W Atlantic; tropical

Lagenorhynchus; short-beaked dolphins; all temperate and polar seas.
L. acutus   Atlantic white-sided   N Atlantic, Atlantic
                          dolphin   Arctic; cold temperate,
                                                          polar
L. albirostris   White-beaked dolphin   N Atlantic, Atlantic
                                                          Arctic; cold temperate,
                                                          polar
L. australis   Peale's dolphin   Southern S America;
                                                          coastal
L. cruciger   Hourglass dolphin   S Atlantic, S Pacific,
                                                          Southern Oc.; temperate
L. obliquidens   Pacific white-sided dolphin   N Pacific; temperate
L. obscurus   Dusky dolphin   Southern Oc.; inshore

Peponocephala; (*Lagenorhynchus*).
P. electra   Melon-headed whale   Tropical seas

Cephalorhynchus; piebald dolphins; southern temperate seas; coastal.
C. commersoni   Commerson's dolphin   Patagonia, S Georgia,
                                                          Kerguelen
C. eutropia   Black dolphin   Chile
C. heavisidii   Heaviside's dolphin   S Africa
C. hectori   Hector's dolphin   New Zealand

Orcaella

O. brevirostris	Irrawaddy dolphin	Bay of Bengal – N Australia; coastal, large rivers

Pseudorca

P. crassidens	False killer whale	All seas; tropical, temperate

Orcinus

O. glacialis	Southern killer	Southern Ocean (Indian Ocean sector); ref. 7.3
O. orca	Killer whale	All seas; mainly polar, temperate

Grampus; (*Grampidelphis*).

G. griseus (*rectipinna*)	Risso's dolphin	All seas; mainly temperate

Globicephala; pilot whales, blackfish.

G. macrorhynchus (*scammonii* ?)	Short-finned pilot whale	Subtropical Atlantic, Pacific
G. melaena (*edwardi*)	Long-finned pilot whale	N Atlantic, southern Oc.; temperate

Feresa

F. attenuata (*occulta*)	Pygmy killer whale (Slender blackfish)	Atlantic, Pacific; tropical, warm temperate

Family Phocoenidae

Porpoises; *c*. 6 species; S temperate, N temperate and Arctic coasts; mainly fisheaters; *.

Phocoena

P. dioptrica	Spectacled porpoise	La Plata – S Georgia, ? New Zealand
P. phocoena (*vomerina*)	Common porpoise (Harbour porpoise)	N Atlantic, N Pacific, Arctic Oc., Black Sea
P. sinus	Gulf porpoise (Cochito)	Gulf of California
P. spinipinnis	Burmeister's porpoise	La Plata – Peru

Phocoenoides

P. dalli (*truei*)	Dall's porpoise	N Pacific; temperate

Neophocaena

N. phocaenoides (*asiaeorientalis*) (*sunameri*)	Finless porpoise	Iran – Borneo – Japan; Yangtzekiang; mainly coastal

Family Monodontidae

White whales; 2 species; Arctic Ocean; *.

Delphinapterus

D. leucas	White whale (Beluga)	Arctic Ocean

Monodon

M. monoceros	Narwhal	Arctic Ocean

Family Physeteridae

Sperm whales; 3 species; all oceans; squid-eaters; *.

Kogia

K. breviceps	Pygmy sperm whale	Tropical, warm temperate seas
K. simus	Dwarf sperm whale	Tropical, subtropical

Physeter

P. catodon (*macrocephalus*)	Sperm whale	All oceans; tropical, (temperate)

Family Ziphiidae

Beaked whales; *c.* 18 species; all oceans; mainly squid-eaters; *.

Tasmacetus

T. shepherdi	Shepherd's beaked whale (Tasman whale)	New Zealand, S America

Berardius

B. arnuxii	Arnoux's beaked whale (Southern four-toothed whale)	Southern hemisphere; temperate
B. bairdii	Baird's beaked whale (Northern four-toothed whale)	N Pacific; temperate

Indopacetus; (*Mesoplodon*); known only from two skulls.

I. pacificus	Indo-pacific beaked whale	Queensland; Somalia

Mesoplodon; (*Micropteron*).

M. bidens	Sowerby's beaked whale	N Atlantic, Baltic; temperate
M. bowdoini	Andrew's beaked whale	SW Pacific, Indian Oc.
M. carlhubbsi	Hubb's beaked whale	N Pacific; warm temperate
M. densirostris	Blainville's beaked whale	All tropical & warm temperate seas
M. europaeus (*gervaisi*)	Gervais' beaked whale	N Atlantic, mainly western; subtropical, warm temperate
M. ginkgodens	Ginkgo-toothed beaked whale	N Pacific, Indian Oc.
M grayi	Gray's beaked whale	S Pacific, S Indian Oc., S Atlantic, (NE Atlantic); temperate
M. hectori	Hector's beaked whale	S Hemisphere, N Pacific; temperate seas
M. layardii	Strap-toothed whale	S Hemisphere; temperate seas
M. mirus	True's beaked whale	N Atlantic; S Africa; temperate
M. stejnegeri	Stejneger's beaked whale	N Pacific; temperate

Ziphius

Z. cavirostris	Cuvier's beaked whale (Goose-beaked whale)	All oceans; temperate, tropical

Hyperoodon, bottle-nose whales.

H. ampullatus	Northern bottlenose whale	Arctic Oc., N Atlantic
H. planifrons	Southern bottlenose whale	S Atlantic, Indian Oc., S Pacific, Southern Oc.

SUBORDER MYSTICETI

Baleen whales; 10 species; all oceans; filter-feeders.

Family Eschrichtiidae

One species; *.

Eschrichtius; (*Rhachianectes*).

E. robustus (*glaucus*) (*gibbosus*)	Grey whale	NE, (NW) Pacific; coastal († N Atlantic)

Family Balaenopteridae

Rorquals; 6 species; all oceans; *.

Balaenoptera; (*Sibbaldus*); rorquals; all oceans; off-shore.

B. acutorostrata	Minke whale (Lesser rorqual)	All oceans
B. borealis	Sei whale	All oceans
B. edeni	Bryde's whale	All oceans
B. musculus (brevicauda)	Blue whale	All oceans
B. physalus	Fin whale (Common rorqual)	All oceans

Megaptera

M. novaeangliae	Humpback whale	All oceans

Family Balaenidae

Right whales; 3 species; temperate, polar seas; mainly coastal; *.

Caperea

C. marginata	Pygmy right whale	S Hemisphere; temperate, polar

Balaena; (*Eubalaena*).

B. glacialis (australis)	Black right whale	All temperate and subarctic seas
B. mysticetus	Bowhead (Greenland right whale)	Arctic Ocean

ORDER SIRENIA

Sea cows; 5 species (one of them extinct); coasts of Indian Ocean, coasts and adjacent rivers of tropical Atlantic Ocean (formerly also N Pacific); herbivores; *.

Family Dugongidae

Dugong

D. dugon	Dugong	E Africa, Red Sea – N Australia; coastal; *

Hydrodamalis

H. gigas †	Steller's sea cow	Bering Sea, N Pacific; extinct

Family Trichechidae

Manatees; 3 species; coasts of tropical Atlantic and adjacent rivers; *.

Trichechus

T. inunguis	Amazon manatee	R Amazon
T. manatus	Caribbean manatee	Virginia – W Indies – Brazil; coastal
T. senegalensis	African manatee	Senegal – Angola; rivers

ORDER PROBOSCIDEA

Family Elephantidae

Elephants; 2 species; Africa, SE Asia; forest, savanna; herbivores; *.

Loxodonta

L. africana (cyclotis)	African elephant	Africa S of Sahara; forest, savanna

Elephas

E. maximus	Indian elephant	India – Sumatra, Sri Lanka; [Borneo]; forest

ORDER PERISSODACTYLA

Odd-toed ungulates; 18 species; Eurasia, Africa, C, S America, herbivores.

Family Equidae

Horses etc.; 9 species; Africa, W, C Asia; steppe, savanna; grazers.

Equus; (Hemionus, Hippotigris).

E. africanus (asinus)	African ass	NE Africa; ancestor of domestic ass E. asinus; *
E. burchellii (quagga)	Common zebra (Burchell's zebra)	E, S Africa; steppe, savanna
E. ferus (caballus) (przewalskii)	Wild horse	Mongolia, Sinkiang; († E Europe); ancestor of domestic horse E. caballus; *
E. grevyi	Grevy's zebra	Kenya, Somalia, S Ethiopia; steppe; *
E. hemionus	Mongolian wild ass	Mongolia, etc.; steppe, subdesert; *

E. kiang (*hemionus*)	Kiang	Tibetan plateau

E. onager	Onager	Turkestan – NW India;
		  († Asia Minor,
		  S Russia, etc.); steppe

E. quagga †	Quagga	S Africa; extinct
E. zebra	Mountain zebra	SW Africa; *

Family Tapiridae

Tapirs; 4 species; C, S America, SE Asia; forest; browsers; *.

Tapirus

T. bairdii	Baird's tapir	S Mexico – Ecuador
T. indicus	Malayan tapir	Burma – Sumatra

T. pinchaque
  (*roulini*)
	Mountain tapir
	  (Woolly tapir)
		Colombia, Ecuador

T. terrestris	Brazilian tapir	Colombia – S Brazil

Family Rhinocerotidae

Rhinoceroses; 5 species; Africa, SE Asia; forest, savanna; browsers, grazers; *.

Rhinoceros; SE Asia; high grass.

R. sondaicus	Javan rhinoceros	Java († Malaya – India)
R. unicornis	Indian rhinoceros	Nepal, NE India

Dicerorhinus; (*Didermoceros*).

D. sumatrensis	Sumatran rhinoceros	Borneo to Malaya,
		  (formerly to Assam);
		  forest

Ceratotherium; (*Diceros*).

C. simum	White rhinoceros
	  (Square-lipped
	  rhinoceros)
		Uganda etc., Zululand;
		  savanna; grazer

Diceros

D. bicornis	Black rhinoceros	Africa S of Sahara; savanna; browser

ORDER HYRACOIDEA

Family Procaviidae

Hyraxes (dassies); *c.* 8 species; Africa, Arabia; forest, rock outcrops in savanna and steppe; herbivores.

Dendrohyrax; tree hyraxes; forest.

D. arboreus	Kenya – S Africa
D. dorsalis	Gambia – Uganda
D. validus	E Tanzania, Zanzibar, etc.

Heterohyrax

H. brucei	Small-toothed rock hyrax	SE Egypt – Transvaal

Procavia; large-toothed rock hyraxes; classification very provisional.

P. capensis	S Africa – S Malawi
(*johnstoni*)	
P. habessinica	Ethiopian Highlands
P. syriaca	Senegal – N Tanzania –
(*ruficeps*)	Egypt, Syria, Arabia
P. welwitschii	N Namibia, S Angola

ORDER TUBULIDENTATA

Family Orycteropidae

One species; *.

Orycteropus

O. afer	Aardvark	Africa S of Sahara;
	(Antbear)	savanna, steppe

ORDER ARTIODACTYLA

Even-toed ungulates; *c.* 192 species; Americas, Eurasia, Africa; herbivores.

Family Suidae

Pigs; 8 species; Eurasia, Africa; forest, savanna; omnivores.

Potamochoerus

P. porcus	Bush pig	Africa S of Sahara,
	(Red river hog)	Madagascar; forest,
		savanna

Sus; Eurasian pigs; Eurasia, (NW Africa); forest.

S. *barbatus*	Bearded pig	Malaya, Sumatra, Borneo, etc.
S. *salvanius*	Pygmy hog	SE Nepal, etc.; *
S. *scrofa*	Wild boar	Europe, NW Africa – SE Siberia – Java, Honshu, Taiwan, Sri Lanka, [New Guinea, New Zealand]; ancestor of domestic pig, S. *domesticus*
S. *verrucosus* (*celebensis*)	Javan pig	Java, Sulawesi, Philippines

Phacochoerus

P. *aethiopicus*	Wart hog	Africa S of Sahara; steppe, savanna

Hylochoerus

H. *meinertzhageni*	Giant forest hog	Liberia – SW Ethiopia; forest

Babyrousa

B. *babyrussa*	Babirusa	Sulawesi, etc.; forest; *

Family Tayassuidae

Peccaries; 3 species; Texas – S America; desert – forest.

Tayassu; (*Dicotyles*).

T. *pecari* (*albirostris*)	White-lipped peccary	Mexico – N Argentina; [Cuba]; forest
T. *tajacu*	Collared peccary	S USA – C Argentina; [Cuba]; desert – forest; (*)

Catagonus

C. *wagneri*	Chaco peccary	Paraguay, N Argentina, Bolivia; grassland, scrub; ref. 8.1; *

Family Hippopotamidae

Hippopotamuses; 2 species; Africa; grazers; *.

Hippopotamus

H. amphibius	Hippopotamus	Africa S of Sahara; grassland with rivers or lakes

Choeropsis

C. liberiensis	Pygmy hippopotamus	Sierra Leone – Nigeria; forest by fresh water

Family Camelidae

Camels, llamas; 4 species; S America, SW, C Asia, ? N Africa; desert, steppe.

Lama

L. guanicoe	Guanaco	Peru – Patagonia; possible ancestor of domestic llama, *L. glama* and alpaca, *L. pacos*; *

Vicugna; (*Lama*)

V. vicugna	Vicugna	Peru – N Chile; montane; *

Camelus

C. ferus (*bactrianus*)	Bactrian camel (Two-humped camel)	C Asia; ancestor of domestic Bactrian camel *C. bactrianus*; *
C. dromedarius	Dromedary (Arabian camel) (One-humped camel)	? Arabia, only known as domestic animal

Family Tragulidae

Chevrotains (mouse-deer); *c.* 4 species; W, C Africa, SE Asia; forest.

Hyemoschus

H. aquaticus	Water chevrotain	Sierra Leone – W Uganda; (*)

Tragulus

T. javanicus	Lesser Malay chevrotain	Indochina – Java, Borneo
T. meminna	Indian spotted chevrotain	India, Sri Lanka
T. napu	Greater Malay chevrotain	S Indochina – Sumatra, Borneo

Family Moschidae

Musk deer; 5 species; E Asia; forest; classification very unstable; *.

Moschus

M. berezovskii		SW China, etc.
M. chrysogaster	Forest musk deer	W Himalayas – Gansu
M. fuscus		Yunnan, China; ref. 8.2
M. moschiferus (sibiricus)	Siberian musk deer	E Siberia
M. sifanicus	Alpine musk deer	W Himalayas – Shaanxi

Family Cervidae

Deer; c. 37 species; Americas, Eurasia, (NW Africa); mainly forest; browsers, grazers.

Hydropotes

H. inermis	Chinese water deer	EC China, Korea, [England]

Muntiacus; muntjacs (barking deer); E Asia.

M. atherodes (pleiharicus)	Bornean yellow muntjac	Borneo; ref. 8.3
M. crinifrons	Black muntjac	Zhejiang, etc., E China; *
M. feae	Fea's muntjac	Thailand; *
M. muntjak	Indian muntjac	India – S China – Java, Borneo, Sri Lanka, Taiwan
M. reevesi	Chinese muntjac	S, E China, Taiwan, [England, France]
M. rooseveltorum	Roosevelts' muntjac	Indochina

Elaphodus

E. cephalophus	Tufted deer	S China, N Burma

Cervus; (Axis, Dama, Sika); Eurasia, (N America).

C. albirostris	Thorold's deer	E Tibetan Plateau; *
C. alfredi		Visayan Is, Philippines; ref. 8.4
C. axis	Spotted deer (Chital)	India etc., Sri Lanka, [New Zealand, Argentina, USA, etc.]

C. dama (mesopotamica)	Fallow deer	S, [W, C] Europe – S Iran, [N, S America, S Africa, Australia, New Zealand, etc.]; (*)
C. duvaucelii	Swamp deer (Barasingha)	India, SW Nepal; *
C. elaphus (canadensis)	Red deer, Wapiti (American elk)	W Europe, NW Africa – W China; W Canada, W USA; [New Zealand, Argentina]; (*)
C. eldii	Thamin	Assam – Indochina, Hainan; *
C. mariannus (unicolor)		Luzon, Mindanao, Mindoro, etc., Philippines; ref. 8.4
C. nippon	Sika deer	SE Siberia – E China, Japan, Taiwan, [W Europe, New Zealand]; (*)
C. porcinus (calamianensis) (kuhli)	Hog-deer	NW India – Indochina, Bawean I (Java), Calamian Is (Philippines), [Australia, Sri Lanka]; (*)
C. schomburgki †	Schomburgk's deer	Thailand; extinct
C. timorensis	Timor deer (Rusa deer)	Java, Sulawesi, Timor, etc., [N Australia, New Zealand]
C. unicolor	Sambar	India – S China – Java, Borneo, Sulawesi, Philippines, Sri Lanka, [Australia, New Zealand]

Elaphurus

E. davidianus	Père David's deer	NE China; extinct in wild

Alces

A. alces	Moose, Elk (Europe)	Scandinavia – E Siberia; Alaska, Canada, N USA; [New Zealand]; con. forest

Rangifer

R. tarandus (caribou)	Reindeer, Caribou	Scandinavia – E Siberia; Alaska, Canada, Greenland; tundra; includes domestic reindeer

Odocoileus; (Blastocerus, Ozotoceros).

O. bezoarticus (campestris)	Pampas deer	E Brazil – N Argentina; grassland; *
O. hemionus	Mule deer (Black-tailed deer)	S Alaska – W, C USA – N Mexico; (*)
O. virginianus	White-tailed deer	S Canada – Peru, N Brazil; [Cuba, etc., New Zealand]; (*)

Blastocerus; (Odocoileus).

B. dichotomus	Marsh deer	S Brazil – NE Argentina; *

Hippocamelus; guemals (huemuls); Andes; *.

H. antisensis	Peruvian guemal	Ecuador – NW Argentina
H. bisulcus	Chilean guemal	S, C Chile, SW Argentina

Mazama; brockets; C, S America.

M. americana	Red brocket	C Mexico – N Argentina; (*)
M. chunyi	Dwarf brocket	S Peru, N Bolivia
M. gouazoubira	Brown brocket	S Mexico – N Argentina
M. rufina	Little red brocket	Venezuela – Ecuador

Pudu; pudus; Andes; *.

P. mephistophiles	Northern pudu	Colombia – N Peru
P. pudu	Southern pudu	S Chile, SW Argentina

Capreolus

C. capreolus (pygargus)	Roe deer	W Europe – SE Siberia, S China

Family Giraffidae

Giraffe, Okapi; 2 species, Africa; browsers.

Okapia

O. johnstoni	Okapi	Zaire; forest

Giraffa
G. camelopardalis   Giraffe   Africa S of Sahara;
                                                  savanna

Family Antilocapridae

One species; sometimes included in the Bovidae.

Antilocapra
A. americana   Pronghorn   W, C USA – N Mexico;
                                                  [Argentina]; desert,
                                                  grassland; (*)

Family Bovidae

Cattle, antelope, sheep, goats; *c.* 126 species; N America, Eurasia, Africa; grazers, browsers.

Tragelaphus; (*Limnotragus, Boocercus, Strepsiceros, Taurotragus*).
T. angasii   Nyala   Natal – Malawi; savanna
T. buxtoni   Mountain nyala   Ethiopia; montane forest,
                                                  grassland
T. derbianus   Giant eland   Senegal – Nile;
                                                  savanna; (*)
T. eurycerus   Bongo   Sierra Leone – Kenya;
                                                  forest; (*)
T. imberbis   Lesser kudu   Ethiopia – Tanzania,
                                                  ? S Arabia; steppe
T. oryx   Eland   E, S Africa; savanna
T. scriptus   Bushbuck   Africa S of Sahara; forest,
                                                  scrub
T. spekii   Sitatunga   Africa S of Sahara;
                                                  swamps; (*)
T. strepsiceros Greater kudu   Chad – Ethiopia –
                                                  S Africa; savanna

Boselaphus
B. tragocamelus Nilgai   India; forest, scrub

Tetracerus
T. quadricornis Four-horned antelope   India; woodland; (*)

Bubalus; (*Anoa*); Asiatic buffaloes; SE Asia; *.
B. arnee   Water buffalo   India; ancestor of domestic
  (*bubalis*)   *B. bubalis*
B. depressicornis Lowland anoa   Sulawesi; forest

B. mindorensis	Tamarau	Mindoro, Philippines; swamp, forest
B. quarlesi	Mountain anoa	Sulawesi

Bos; (*Bibos*); oxen; Eurasia; forest; *.

B. gaurus	Gaur (Indian bison)	India – Malaya; forest; ancestor of domestic mithan (gayal), *B. frontalis*
B. javanicus (*banteng*)	Banteng	Burma – Java, Borneo; forest; includes domestic Bali cattle
B. mutus (*grunniens*)	Yak	Tibetan Plateau; montane grassland; ancestor of domestic yak, *B. grunniens*
B. primigenius (*taurus*)	Aurochs (Urus)	Europe, W Asia; extinct in wild but ancestor of domestic cattle, *B. taurus*, including zebu, *B. indicus*
B. sauveli	Kouprey	Indochina; forest

Synceros

S. caffer (*nanus*)	African buffalo	Africa S of Sahara; forest, savanna

Bison

B. bison	American bison	N America; grassland (woodland); (*)
B. bonasus	European bison (Wisent)	E Europe; forest; *

Cephalophus; (*Philantomba*); forest duikers; Africa S of Sahara; forest.

C. adersi	Aders' duiker	Zanzibar, E Kenya
C. callipygus	Peters' duiker	Gabon, Cameroun, etc.
C. dorsalis	Bay duiker	Guinea – Zaire, etc.
C. jentinki	Jentink's duiker	Liberia, etc.; *
C. leucogaster	White-bellied duiker	Cameroun – E Zaire
C. maxwellii	Maxwell's duiker	Senegal – Nigeria
C. monticola	Blue duiker	SE Nigeria – Kenya – S Africa; *
C. natalensis	Red forest duiker	Somalia – E Zaire – S Africa
C. niger	Black duiker	Guinea – Nigeria

C. nigrifrons	Black-fronted duiker	Cameroun – Kenya – Angola; wet forest
C. ogilbyi	Ogilby's duiker	Sierra Leone – Gabon
C. rufilatus	Red-flanked duiker	Senegal – Sudan; forest-edge, thicket
C. spadix	Abbott's duiker	Tanzania; montane
C. sylvicultor	Yellow-backed duiker	Gambia – Kenya – Angola
C. weynsi		Zaire – W Kenya
C. zebra	Banded duiker	Liberia, etc.

Sylvicapra

S. grimmia	Common duiker	Africa S of Sahara; savanna

Kobus; (*Adenota, Onotragus*); Africa S of Sahara.

K. ellipsiprymnus (*defassa*)	Waterbuck	Africa S of Sahara; savanna
K. kob	Kob	Volta – Kenya; savanna
K. leche	Lechwe	Zambia, etc.; wet grassland; *
K. megaceros	Nile lechwe	S Sudan; swamps
K. vardonii	Puku	Zambia, etc.; savanna

Redunca; reedbuck; Africa S of Sahara; savanna.

R. arundinum	Reedbuck	S Africa – L Victoria
R. fulvorufula	Mountain reedbuck	S, E, C Africa
R. redunca	Bohar reedbuck	Senegal – Tanzania

Pelea

P. capreolus	Rhebok	S Africa; hilly grassland

Hippotragus

H. equinus	Roan antelope	Africa S of Sahara; savanna; *
H. leucophaeus †	Blue buck	SW Africa; extinct
H. niger	Sable antelope	S Africa – Kenya; savanna; (*)

Oryx

O. dammah (*tao*)	Scimitar oryx	S edge of Sahara; semi-desert; *
O. gazella (*beisa*)	Gemsbok (Beisa)	SW, E Africa; steppe
O. leucoryx	Arabian oryx	Arabia; desert; *

Addax

A. nasomaculatus	Addax	Sahara; desert, semi-desert; *

Connochaetes; (*Gorgon*); wildebeest; steppe.

C. gnou	Black wildebeest (White-tailed gnu)	S Africa; *
C. taurinus	Blue wildebeest (Brindled gnu)	S, E Africa

Alcelaphus

A. buselaphus	Red hartebeest	W, E, SW Africa; steppe; (*)

Sigmoceros

S. lichtensteinii	Lichtenstein's hartebeest	Angola – Tanzania; savanna

Damaliscus; (*Beatragus*).

D. dorcas	Bontebok, Blesbok	S Africa; steppe; (*)
D. hunteri	Hunter's hartebeest	NE Kenya, S Somalia; steppe; *
D. lunatus (*korrigum*)	Tsessebi, Topi	Africa S of Sahara; savanna; (*)

Oreotragus

O. oreotragus	Klipspringer	Africa S of Sahara; rocky hills in savanna

Madoqua; (*Rhynchotragus*); dik-diks; E, S Africa; dry steppe; ref. 8.5.

M. guentheri	Gunther's dik-dik	E Africa
M. kirkii	Kirk's dik-dik (Damara dik-dik)	SW, E Africa
M. piacentinii		SE Somalia
M. saltiana (*phillipsi*) (*swaynei*)	Salt's dik-dik (Swayne's dik-dik)	N, E Ethiopia, Somalia

Dorcatragus

D. megalotis	Beira antelope	Somalia, E Ethiopia; semi-desert; *

Ourebia

O. ourebi	Oribi	Africa S of Sahara; steppe, savanna

Raphiceros

R. campestris	Steenbok	S, E Africa; steppe, savanna
R. melanotis	Cape grysbok	SW Africa; steppe
R. sharpei	Sharpe's grysbok	S Africa – Tanzania; savanna

Neotragus; (*Nesotragus*).

N. batesi	Bates' dwarf antelope	SE Nigeria – E Zaire, etc.; forest
N. moschatus	Suni	Kenya – Natal; thicket; (*)
N. pygmaeus	Royal antelope	Sierra Leone – Ghana; forest

Aepyceros

A. melampus	Impala	S, E Africa; savanna; (*)

Antilope

A. cervicapra	Blackbuck	India; [Texas]; steppe; (*)

Antidorcas

A. marsupialis	Springbok	S Africa; steppe

Litocranius

L. walleri	Gerenuk	E Africa; steppe

Ammodorcas

A. clarkei	Dibatag	Somalia, E Ethiopia; semi-desert; *

Gazella; gazelles; Sahara – Tanzania – Mongolia; desert, steppe.

G. arabica		Farsan I, Red Sea
G. bilkis		Yemen; ref. 8.7
G. cuvieri	Edmi gazelle	Morocco, N Algeria, Tunis; *
G. dama	Addra gazelle	W, S Sahara; *
G. dorcas	Dorcas gazelle	Sahara – C India; (*)
G. gazella	Mountain gazelle (Idmi)	Arabia – Israel; (*)
G. granti	Grant's gazelle	Ethiopia – Tanzania
G. leptoceros	Sand gazelle (Rhim)	N Sahara; *
G. rufifrons	Red-fronted gazelle	Senegal – Ethiopia
G. rufina †	Red gazelle	Algeria; extinct
G. soemmerringii	Soemmerring's gazelle	Sudan, Ethiopia, Somalia

G. spekei	Speke's gazelle (Dero)	Somalia, E Ethiopia; *
G. subgutturosa	Goitred gazelle	Arabia – Pakistan – Mongolia; (*)
G. thomsonii	Thomson's gazelle	Sudan – Tanzania

Procapra; Chinese gazelles; Mongolia – Tibet; steppe.

P. gutturosa	Mongolian gazelle	Mongolia, etc.
P. picticaudata	Tibetan gazelle	Tibetan Plateau
P. przewalskii	Przewalski's gazelle	C China

Pantholops

P. hodgsonii	Chiru (Tibetan antelope)	Tibetan Plateau; *

Saiga

S. tatarica	Saiga	S Russia – Mongolia; steppe

Nemorhaedus; gorals; E Asia; montane forest.

N. cranbrooki	Red goral	N Burma
N. goral	Common goral	Himalayas – SE Siberia; *

Capricornis; serows; E, SE Asia; forest.

C. crispus	Japanese serow	Japan, Taiwan
C. sumatraensis	Mainland serow	Kashmir – C China – Sumatra; *

Oreamnos

O. americanus	Mountain goat	NW USA – S Alaska

Rupicapra

R. rupicapra	Chamois	N Spain – Caucasus, [New Zealand], montane; (*)

Ovibos

O. moschatus	Musk ox	Alaska – Greenland; [Norway, Spitzbergen, N Siberia]; tundra

Budorcas

B. taxicolor	Takin	E Himalayas, SW China; montane forest; (*)

Hemitragus; tahrs; SW, S Asia; montane.

H. hylocrius	Nilgiri tahr	S India; *
H. jayakari	Arabian tahr	Oman; *
H. jemlahicus	Himalayan tahr	Himalayas, [New Zealand, California, S Africa]

Capra; goats; S Europe – C Asia – NE Africa; montane.

C. aegagrus (*hircus*)	Wild goat (Bezoar)	Pakistan – Asia Minor, Crete; ancestor of domestic goat *C. hircus*
C. caucasica	West Caucasian tur	W Caucasus
C. cylindricornis	East Caucasian tur	E Caucasus
C. falconeri	Markhor	W Himalayas, Afghanistan, etc.; *
C. ibex (*sibirica*) (*walie*)	Ibex	Alps – C Asia – Ethiopia; (*)
C. pyrenaica	Spanish ibex	Spain; (*)

Ammotragus; (*Capra*).

A. lervia	Barbary sheep (Aoudad)	Sahara, etc., [S, SW USA]; (*)

Pseudois

P. nayaur	Bharal (Blue sheep)	Himalayas, W China; montane
P. schaeferi	Dwarf blue sheep	Upper Yangtze, China

Ovis; sheep; W, C, NE Asia, W N America; montane; classification unstable.

O. ammon	Argali	Altai – Himalayas; *
O. canadensis	American bighorn (Mountain sheep)	SW Canada – W USA – N Mexico; *
O. dalli	Dall sheep (White sheep)	Alaska – N Br. Colombia
O. nivicola	Siberian bighorn	NE Siberia
O. orientalis (*laristanica*) (*musimon*)	Mouflon	Iran – Asia Minor, [Sardinia, Corsica, Cyprus, C Europe]; ancestor of domestic sheep, *O. aries*; (*)
O. vignei	Urial	Kashmir – Iran, Turkestan; *

ORDER PHOLIDOTA

One family.

Family Manidae

Pangolins (scaly anteaters); 7 species; Africa, S Asia; forest, savanna; arboreal, terrestrial; feed on ants and termites; ref. 8.6; *.

Manis; (*Paramanis*); Asian pangolins.

M. crassicaudata	Indian pangolin	India, Sri Lanka
M. javanica	Malayan pangolin	Burma – Java, Borneo, Palawan
M. pentadactyla	Chinese pangolin	Nepal – S China, Taiwan

Phataginus; (*Smutsia*, *Uromanis*); African pangolins.

P. gigantea	Giant ground pangolin	Senegal – Uganda – Angola; savanna, forest
P. temmincki	Temminck's ground pangolin	S Sudan – S Africa; savanna
P. tetradactyla (*longicaudata*)	Long-tailed pangolin	Senegal – Uganda – Angola; forest, arboreal
P. tricuspis	Tree pangolin	Senegal – W Kenya – Angola; forest, arboreal

ORDER RODENTIA

Rodents; *c.* 1738 species; worldwide; terrestrial, arboreal, subterranean, aquatic; mainly seed-eaters, also herbivores and insectivores.

Family Aplodontidae

One species.

Aplodontia

A. rufa	Mountain beaver	NW USA; wet forest

Family Sciuridae

Squirrels; *c.* 255 species; Eurasia, Africa, N, C, S America; forest, savanna, (steppe).

Subfamily Sciurinae

Tree and ground squirrels; *c.* 217 species.

Sciurus; (*Guerlinguetus*, *Neosciurus*); Palaearctic and American tree squirrels; forest; arboreal.

S. aberti (*kaibabensis*)	Tassel-eared squirrel	Colorado – NW Mexico, Arizona; montane pine forest
S. aestuans (*gilvigularis*)		N Argentina – Venezuela
S. alleni	Allen's squirrel	NE Mexico
S. anomalus	Persian squirrel	Iran, Asia Minor, etc.
S. arizonensis	Arizona grey squirrel	Arizona, N Mexico; montane
S. aureogaster (*griseoflavus*) (*nelsoni*) (*poliopus*)	Mexican grey squirrel	C Mexico – Guatemala
S. carolinensis	Eastern grey squirrel	E, C USA, SE Canada, [Britain, Ireland, S Africa]; dec. forest
S. colliaei	Collie's squirrel	NW Mexico, montane
S. deppei	Deppe's squirrel	E Mexico – Costa Rica
S. flammifer		Venezuela; (*)
S. granatensis	Tropical red squirrel	Ecuador, Venezuela – Costa Rica
S. griseus	Western grey squirrel	Washington – California
S. ignitus		N Argentina – Peru, W Brazil
S. igniventris		Colombia – Peru, N Brazil
S. lis	Japanese squirrel	Japan, except Hokkaido
S. nayaritensis (*apache*)	Nayarit squirrel	NW Mexico – S Arizona
S. niger	Eastern fox squirrel	E, C USA; open forest; (*)
S. oculatus	Peters' squirrel	C Mexico
S. pucheranii		Colombia
S. pyrrhinus		E Peru
S. richmondi	Richmond's squirrel	Nicaragua
S. sanborni		SE Peru
S. spadiceus (*langsdorffi*) (*pyrrhonotus*)		Amazon Basin
S. stramineus		NE Peru, SE Ecuador
S. variegatoides (*goldmani*)	Variegated squirrel	SE Mexico – Panama
S. vulgaris	Eurasian red squirrel	Ireland – Hokkaido; con. forest
S. yucatanensis	Yucatan squirrel	Yucatan, Guatemala

Syntheosciurus; Central America; montane forest.

S. brochus (*poasensis*)	Panama mountain squirrel	Panama, Costa Rica

Microsciurus; American pygmy squirrels.

M. alfari	Alfaro's pygmy squirrel	S Nicaragua – Panama
M. flaviventer		N Brazil, Peru
M. mimulus (*isthmius*) (*boquetensis*)	Cloud-forest pygmy squirrel	Panama – Ecuador
M. santanderensis		Colombia

Sciurillus

S. pusillus	South American pygmy squirrel	S America N of Amazon; forest

Prosciurillus; Sulawesi dwarf squirrels; Sulawesi.

P. abstrusus		Sulawesi
P. leucomus		Sulawesi
P. murinus		Sulawesi

Rheithrosciurus

R. macrotis	Tufted ground squirrel	Borneo; forest

Tamiasciurus; American red squirrels (chickarees); N America; coniferous forest.

T. douglasii	Douglas' squirrel	W coast USA
T. hudsonicus (*fremonti*)	American red squirrel	Canada, NE USA, Rockies
T. mearnsi		Baja California

Funambulus; palm squirrels (Indian striped squirrels); India etc.; forest, scrub; terrestrial.

F. layardi	Layard's striped squirrel	Sri Lanka, S India
F. palmarum	Indian palm squirrel	Sri Lanka, India
F. pennantii	Northern palm squirrel	N, C India, Pakistan, Nepal, [SE, SW Australia]
F. sublineatus	Dusky striped squirrel	Sri Lanka, S India
F. tristriatus	Jungle striped squirrel	S India

Ratufa; oriental giant squirrels; SE Asia; forest; *.

R. affinis	Cream-coloured giant squirrel	Malaya, Sumatra, Borneo
R. bicolor	Black giant squirrel	Nepal – S China – Java
R. indica	Indian giant squirrel	S, C India
R. macroura	Grizzled Indian squirrel	S India, Sri Lanka

Protoxerus; African giant squirrels; W, C, E Africa; forest.

P. aubinni	Slender-tailed giant squirrel	Liberia – Ghana
P. stangeri	Giant forest squirrel	Sierra Leone – Kenya – Angola

Epixerus; African palm squirrels; high forest; *.

E. ebii	Temminck's giant squirrel	Sierra Leone – Ghana
E. wilsoni		Cameroun – R Zaire

Funisciurus; African striped squirrels, rope squirrels; Africa; forest.

F. anerythrus	Thomas' tree squirrel	Senegal, Nigeria – Uganda, Zaire
F. bayonii	Bocage's tree squirrel	NE Angola, SW Zaire
F. carruthersi	Mountain tree squirrel	Ruwenzori – Burundi; montane
F. congicus	Kuhl's tree squirrel	R Zaire – SW Africa
F. isabella	Gray's four-striped squirrel	Cameroun – R Zaire
F. lemniscatus	Leconte's four-striped squirrel	R Sanaga – R Zaire
F. leucogenys	Orange-headed squirrel	Ghana – Cent. Af. Rep. – Rio Muni
F. pyrrhopus	Cuvier's tree squirrel	Gambia – Uganda – Angola
F. substriatus	De Winton's tree squirrel	Ivory Coast – Nigeria

Paraxerus; (*Aethosciurus*, *Montisciurus*); African bush squirrels; Africa; savanna, dry forest.

P. alexandri	Alexander's bush squirrel	NE Zaire, Uganda
P. boehmi	Boehm's bush squirrel	E, C Africa
P. cepapi	Smith's bush squirrel (S African tree squirrel)	S Angola – S Tanzania – Transvaal
P. cooperi	Cooper's green squirrel	Cameroun
P. flavivittis	Striped bush squirrel	N Mozambique – S Kenya
P. lucifer	Black and red bush squirrel	N Malawi, SW Tanzania; montane
P. ochraceus	Huet's bush squirrel	Tanzania – S Sudan
P. palliatus	Red bush squirrel	Natal – Somalia
P. poensis	Small green squirrel	Sierra Leone – R Zaire
P. vexillarius	Swynnerton's bush squirrel	C, E Tanzania; montane
P. vincenti	Vincent's bush squirrel	N Mozambique; (in *P. vexillarius*?)

Heliosciurus; sun squirrels; Africa; forest; arboreal.

H. gambianus	Gambian sun squirrel	Senegal – Ethiopia – Zambia; savanna, secondary forest
H. rufobrachium	Red-legged sun squirrel	Senegal – Kenya – Zimbabwe; forest
H. ruwenzorii	Ruwenzori sun squirrel	E Zaire, etc.; montane forest

Hyosciurus

H. heinrichi	Sulawesi long-nosed squirrel	Sulawesi; montane forest

Myosciurus

M. pumilio	African pygmy squirrel	SE Nigeria – Gabon; forest

Callosciurus; (*Rubrisciurus*); oriental tree squirrels; SE Asia.

C. adamsi	Ear-spot squirrel	N, NW Borneo
C. albescens		Sumatra
C. baluensis	Kinabalu squirrel	C, N Borneo; montane
C. caniceps	Grey-bellied squirrel (Golden-backed squirrel)	E Himalayas – Malaya, Taiwan
C. erythraeus (*flavimanus*)	Belly-banded squirrel	S China – Malaya, Taiwan
C. finlaysonii (*ferrugineus*)	Finlayson's squirrel (Variable squirrel)	E Burma – Indochina
C. inornatus		Indochina
C. melanogaster		Mentawai Is, Sumatra
C. nigrovittatus	Black-banded squirrel	Malaya, Sumatra, Java, Borneo
C. notatus	Plantain squirrel	Malaya – Java, Borneo; secondary forest
C. phayrei		Burma
C. prevostii	Prevost's squirrel	Malaya, Sumatra, Borneo, Sulawesi
C. pygerythrus	Irrawaddy squirrel	Nepal – Indochina
C. quinquestriatus	Anderson's squirrel	S China, Burma
C. rubriventer		Sulawesi

Tamiops; (*Callosciurus*); Oriental striped squirrels; S China, etc.

T. macclellandi	Himalayan striped squirrel	Nepal – S China – Malaya
T. maritimus		S China, Indochina, Taiwan
T. rodolphei	Cambodian striped squirrel	Indochina
T. swinhoei	Swinhoe's striped squirrel	NE China – N Vietnam

Sundasciurus; (*Callosciurus*); Sunda squirrels; SE Asia; forest; inclusion of species in this genus very provisional.

S. brookei	Brooke's squirrel	C, N Borneo
S. hippurus	Horse-tailed squirrel	Malaya, Sumatra, Borneo
S. hoogstraali		Busuanga, Philippines
S. jentinki	Jentink's squirrel	Borneo
S. juvencus		Palawan, Philippines
S. lowii	Low's squirrel	Malaya, Sumatra, Borneo
S. mindanensis		Mindanao, Philippines
S. mollendorffi (*albicauda*)		Calamian Is, Philippines
S. philippinensis		Basilan, Mindanao, Philippines
S. rabori		Palawan
S. samarensis		Samar, Philippines
S. steerii		Palawan, Philippines
S. tenuis	Slender squirrel	Malaya, Sumatra, Borneo

Menetes

M. berdmorei	Berdmore's squirrel (Indochinese ground squirrel)	Burma – Indochina

Rhinosciurus

R. laticaudatus	Shrew-faced squirrel (Long-nosed squirrel)	Malaya, Sumatra, Borneo; forest

Lariscus

L. hosei	Four-striped ground squirrel	N, NW Borneo; forest; *.
L. insignis	Three-striped ground squirrel	Malaya – Java, Borneo; forest
L. niobe		Mentawei Is, Sumatra

Dremomys

D. everetti	Bornean mountain ground squirrel	N, NW Borneo; montane forest
D. lokriah	Orange-bellied Himalayan squirrel	Nepal – N Burma; montane forest
D. pernyi	Perny's long-nosed squirrel	S China – Burma, Taiwan
D. pyrrhomeris		C, S China
D. rufigenis	Red-cheeked squirrel	Indochina – Malaya

Sciurotamias
S. davidianus	Père David's rock squirrel	Hebei – Sichuan, China
S. forresti	Forrest's rock squirrel	Yunnan, China

Glyphotes
G. canalvus	Grey-bellied sculptor squirrel	Mt Dulit, Borneo
G. simus	Red-bellied sculptor squirrel	Mt Kinabalu, Borneo

Nannosciurus
N. melanotis	Black-eared pygmy squirrel	Sumatra, Java, Borneo

Exilisciurus
E. concinnus		Basilan, Philippines
E. exilis	Plain pygmy squirrel	Borneo; lowland forest
E. luncefordi		Mindanao, Philippines
E. samaricus		Samar, Philippines
E. surrutilus		Mindanao, Philippines
E. whiteheadi	Whitehead's pygmy squirrel	NW Borneo; montane forest

Atlantoxerus
A. getulus	Barbary ground squirrel	Morocco, Algeria

Xerus; (*Euxerus, Geosciurus*); African ground squirrels; Africa S of Sahara; savanna, steppe; terrestrial.
X. erythropus	Geoffroy's ground squirrel	Morocco, Senegal – Kenya, Ethiopia
X. inauris	Cape ground squirrel	Africa S of Zambezi
X. princeps	Kaokoveld ground squirrel (Mountain ground squirrel)	S W Africa, S Angola
X. rutilus	Unstriped ground squirrel	Ethiopia – N Tanzania

Spermophilopsis
S. leptodactylus	Long-clawed ground squirrel	Russian Turkestan etc.; desert

Marmota; marmots; Palaearctic, N America; alpine talus, grassland, (forest).
M. baibacina	Steppe marmot	Kazakhstan, etc.
M. bobak	Bobak marmot	S Russia
M. broweri	Alaska marmot	N Alaska; vicariant of *M. camtschatica*
M. caligata	Hoary marmot	Alaska – Idaho; montane

M. camtschatica	Black-capped marmot	NE Siberia; montane
M. caudata	Long-tailed marmot	Tien Shan – Kashmir; montane
M. flaviventris	Yellow-bellied marmot	W USA; montane
M. himalayana	Himalayan marmot	Himalayas
M. marmota	Alpine marmot	C Europe; montane
M. menzbieri	Menzbier's marmot	W Tien Shan; *
M. monax	Woodchuck	Alaska – Labrador, E USA; forest
M. olympus	Olympic marmot	Olympic Peninsula, Washington, USA
M. sibirica	Siberian marmot	SE Siberia, N Mongolia, etc.
M. vancouverensis	Vancouver marmot	Vancouver I, Canada; *

Cynomys; prairie dogs; C, W USA – C Mexico; grassland.

C. gunnisoni	Gunnison's prairie dog	Colorado – Arizona; high grassland
C. leucurus	White-tailed prairie dog	Wyoming, etc.
C. ludovicianus	Black-tailed prairie dog	C USA; grassland
C. mexicanus	Mexican prairie dog	C Mexico; *
C. parvidens	Utah prairie dog	Utah; *

Spermophilus; (*Callospermophilus*, *Citellus*, *Otospermophilus*); ground squirrels, sousliks; N America, N Eurasia; grassland, steppe, (open forest).

S. adocetus	Tropical ground squirrel	C Mexico
S. alaschanicus	Alashan ground squirrel	W Gansu, S Mongolia
S. annulatus	Ring-tailed ground squirrel	W Mexico
S. armatus	Uinta ground squirrel	Rocky Mts (USA)
S. atricapillus	Baja California rock squirrel	Baja California
S. beecheyi	California ground squirrel	California, W Oregon
S. beldingi	Belding's ground squirrel	W USA
S. brunneus	Idaho ground squirrel	W Idaho
S. citellus	European souslik	SE Europe
S. columbianus	Columbian ground squirrel	NW USA, SW Canada
S. dauricus	Daurian ground squirrel	E Mongolia, Transbaikalia, N China
S. elegans (*richardsonii*)		C Rockies of USA
S. erythrogenys (*major*)		E Kazakhstan – Mongolia
S. franklinii	Franklin's ground squirrel	C USA, SC Canada; tall grass

S. fulvus	Large-toothed souslik	Turkestan
S. lateralis	Golden-mantled ground squirrel	W USA, SW Canada; montane forest, scrub
S. madrensis	Sierra Madre ground squirrel	Chihuahua, Mexico
S. major	Russet souslik	S Russia – SW Siberia
S. mexicanus	Mexican ground squirrel	E Mexico, Texas
S. mohavensis	Mohave ground squirrel	Mohave Desert, S California
S. musicus (citellus)		N Caucasus
S. parryii	Arctic souslik (Arctic ground squirrel)	NE Siberia; Alaska – Hudson Bay; tundra
S. perotensis	Perote ground squirrel	Veracruz, Mexico
S. pygmaeus	Little souslik	S Russia, Kazakhstan
S. relictus	Tien Shan souslik	Tien Shan Mts
S. richardsoni	Richardson's ground squirrel	NW USA, SW Canada; grassland
S. saturatus	Cascade golden-mantled ground squirrel	Cascade Mts, Washington, Br. Columbia
S. spilosoma	Spotted ground squirrel	SC USA, N Mexico
S. suslicus	Spotted souslik	E Europe, S Russia
S. tereticaudus	Round-tailed ground squirrel	SW USA, NW Mexico; desert, scrub
S. townsendii	Townsend's ground squirrel	W USA; steppe
S. tridecemlineatus	Thirteen-lined ground squirrel	C USA, SC Canada; short grass
S. undulatus	Long-tailed souslik	Tien Shan – R Lena
S. variegatus	Rock squirrel	SW USA, Mexico; dry steppe, desert
S. washingtoni	Washington ground squirrel	Washington, Oregon
S.xanthoprymnus (citellus)		Asia Minor, etc.

Ammospermophilus; antelope-squirrels; SW USA, N Mexico; desert, steppe.

A. harrisii	Harris' antelope-squirrel	Arizona, N Mexico
A. interpres	Texas antelope-squirrel	New Mexico – N Mexico
A. leucurus (insularis)	White-tailed antelope-squirrel	SW USA
A. nelsoni	Nelson's antelope-squirrel	California

Tamias; (Eutamias); chipmunks; N America, N Eurasia.

T. alpinus	Alpine chipmunk	California; high montane
T. amoenus	Yellow-pine chipmunk	NW USA, SW Canada

T. bulleri	Buller's chipmunk	N Mexico
T. canipes	Grey-footed chipmunk	New Mexico, Texas
T. cinereicollis	Grey-collared chipmunk	Arizona, New Mexico; montane forest
T. dorsalis	Cliff chipmunk	SW USA, N Mexico; pine, juniper forest
T. merriami	Merriam's chipmunk	S California; scrub, forest
T. minimus	Least chipmunk	W USA, S, C Canada; forest, scrub
T. obscurus (*merriami*)	Chaparral chipmunk (California chipmunk)	S California
T. palmeri	Palmer's chipmunk	Nevada; montane forest
T. panamintinus	Panamint chipmunk	Arizona, California; scrub
T. quadrimaculatus	Long-eared chipmunk	California
T. quadrivittatus	Colorado chipmunk	SW USA; montane forest
T. ruficaudus	Red-tailed chipmunk	NW USA, SW Canada; montane forest
T. rufus		W Colorado, etc.; ref. 9.5
T. sibiricus	Siberian chipmunk	Siberia, W China, Hokkaido
T. sonomae	Sonoma chipmunk	NW California; scrub
T. speciosus	Lodgepole chipmunk	California
T. striatus	Eastern chipmunk	E USA, SE Canada; forest
T. townsendii (*ochrogenys*) (*senex*) (*siskiyou*)	Townsend's chipmunk	W coast USA; ref. 9.6
T. umbrinus	Uinta chipmunk	SW USA; montane forest

Subfamily Petauristinae

Flying squirrels (gliding squirrels); *c.* 38 species; SE Asia, (N Eurasia, N America).

Petaurista; giant flying squirrels; E, SE Asia; forest.

P. alborufus	Red and white flying squirrel	S China – Thailand, Taiwan
P. elegans	Spotted giant flying squirrel	Nepal – Java, Borneo
P. leucogenys	Japanese giant flying squirrel	Japan except Hokkaido; Gansu – Yunnan
P. magnificus	Hodgson's flying squirrel	Nepal, Sikkim, etc.
P. petaurista	Red giant flying squirrel	Kashmir – S China – Java, Borneo, Sri Lanka

Biswamoyopterus
B. biswasi	Namdapha flying squirrel	Arunachal Pradesh, NE India; ref. 9.7

Eupetaurus
E. cinereus	Woolly flying squirrel	Kashmir; montane coniferous forest, rocks

Pteromys
P. momonga	Small Japanese flying squirrel	Honshu, Kyushu, Japan
P. volans	Siberian flying squirrel	Finland – Korea, Hokkaido; coniferous forest

Glaucomys; American flying squirrels; N America; forest.
G. sabrinus	Northern flying squirrel	Canada, W USA
G. volans	Southern flying squirrel	E USA – Honduras

Aeromys
A. tephromelas (*phaeomelas*)	Black flying squirrel	Malaya, Sumatra, Borneo
A. thomasi	Thomas' flying squirrel	Borneo

Hylopetes; pygmy flying squirrels; SE Asia; forest.
H. alboniger	Particoloured flying squirrel	Nepal – Indochina
H. baberi (*fimbriatus*)		Kashmir; montane; ref. 9.8
H. fimbriatus	Kashmir pygmy flying squirrel	Afghanistan – Kashmir
H. electilis	Hainan flying squirrel	Hainan, China; formerly in *Petinomys*
H. lepidus (*spadiceus*)	Red-cheeked flying squirrel	Thailand – Java, Borneo
H. mindanensis		Mindanao, Philippines
H. nigripes		Palawan, Philippines
H. phayrei	Phayre's flying squirrel	Burma, Thailand
H. platyurus	Grey-cheeked flying squirrel	Malaya – Java, Borneo

Petinomys
P. bartelsi		Java
P. crinitus		Basilan, Philippines
P. fuscocapillus	Travancore flying squirrel	S India, Sri Lanka

P. genibarbis	Whiskered flying squirrel	Malaya – Borneo
P. hageni		Borneo, Sumatra
P. sagitta		Java
P. setosus (morrisi)	White-bellied flying squirrel	Burma, Malaya, Sumatra, Borneo
P. vordermanni	Vordermann's flying squirrel	Malaya, Borneo

Aeretes

A. melanopterus	North Chinese flying squirrel	NE China, Sichuan

Trogopterus

T. xanthipes	Complex-toothed flying squirrel	China

Belomys

B. pearsonii	Hairy-footed flying squirrel	Sikkim – Indochina, Taiwan

Pteromyscus

P. pulverulentus	Smoky flying squirrel	S Thailand – Sumatra, Borneo

Petaurillus

P. emiliae		Sarawak, Borneo
P. hosei		Sarawak, Borneo
P. kinlochii	Selangor pygmy flying squirrel	Selangor, Malaya

Iomys

I. horsfieldii	Horsfield's flying squirrel	Malaya – Java, Borneo

Family Geomyidae

Pocket gophers; c. 34 species; USA (especially SW) to Colombia; forest – steppe; subterranean.

Geomys

G. arenarius	Desert pocket gopher	S New Mexico, etc.
G. bursarius	Plains pocket gopher	C USA
G. personatus	Texas pocket gopher	S Texas, NE Mexico
G. pinetis (colonus) (cumberlandius) (fontanelus)	Southeastern pocket gopher	Alabama – C Florida

G. tropicalis (personatus)	Tropical pocket gopher	S Tamaulipas, E Mexico

Thomomys

T. bottae	Botta's pocket gopher	S USA – N Mexico
T. bulbivorus	Camas pocket gopher	NW Oregon
T. clusius		Wyoming
T. idahoensis (talpoides)	Idaho pocket gopher	Idaho, etc.
T. mazama	Western pocket gopher	Oregon, N California
T. monticola	Mountain pocket gopher	California
T. talpoides	Northern pocket gopher	NW USA, SW Canada
T. umbrinus (townsendi)	Southern pocket gopher	SW USA, Mexico

Pappogeomys; (Cratogeomys); Mexico, (S USA).

P. alcorni	Alcorn's pocket gopher	Jalisco, Mexico
P. bulleri	Buller's pocket gopher	SW Mexico
P. castanops	Yellow-faced pocket gopher	SE Colorado – N Mexico
P. fumosus	Smoky pocket gopher	Colima, SW Mexico
P. gymnurus	Llano pocket gopher	C Mexico
P. merriami	Merriam's pocket gopher	C Mexico
P. neglectus	Queretaro pocket gopher	Queretaro, C Mexico
P. tylorhinus	Taylor's pocket gopher	C Mexico
P. zinseri	Zinser's pocket gopher	Jalisco, C Mexico

Orthogeomys; (Heterogeomys, Macrogeomys); Mexico – Colombia.

O. cavator	Chiriqui pocket gopher	W Panama, Costa Rica
O. cherriei	Cherrie's pocket gopher	Costa Rica
O. cuniculus	Oaxacan pocket gopher	Oaxaca, Mexico
O. dariensis	Darien pocket gopher	E Panama – Colombia
O. grandis	Large pocket gopher	Honduras – Guerrero, Mexico
O. heterodus	Variable pocket gopher	Costa Rica
O. hispidus	Hispid pocket gopher	E Mexico – Guatemala
O. lanius	Big pocket gopher	Veracruz, Mexico
O. matagalpae	Nicaraguan pocket gopher	Nicaragua, Honduras
O. pygacanthis	El Salvador pocket gopher	El Salvador
O. underwoodi	Underwood's pocket gopher	Costa Rica

Zygogeomys

Z. trichopus	Michoacan pocket gopher	Michoacan, SW Mexico

Family Heteromyidae

Pocket mice, kangaroo-rats, etc.; *c.* 60 species; N, (C, S) America; desert, steppe, (forest).

Perognathus; pocket mice; Mexico, SW, (NW) USA; desert, steppe.

P. alticola	White-eared pocket mouse	S California
P. amplus	Arizona pocket mouse	Arizona, etc.
P. anthonyi	Anthony's pocket mouse	Cerros I, Baja California
P. arenarius	Little desert pocket mouse	Baja California
P. artus	Narrow-skulled pocket mouse	NW Mexico
P. baileyi	Bailey's pocket mouse	Baja California, Arizona, Sonora
P. californicus	California pocket mouse	California
P. dalquesti	Dalquest's pocket mouse	Baja California
P. fallax	San Diego pocket mouse	S California, N Baja California
P. fasciatus	Olive-backed pocket mouse	Colorado – Saskatchewan
P. flavescens (*apache*)	Plains pocket mouse	Arizona – N Texas – Minnesota
P. flavus (*merriami*)	Silky pocket mouse	W Nebraska – C Mexico
P. formosus	Long-tailed pocket mouse	Utah – Baja California
P. goldmani	Goldman's pocket mouse	NW Mexico
P. hispidus	Hispid pocket mouse	N Dakota – C Mexico
P. inornatus	San Joaquin pocket mouse	California
P. intermedius	Rock pocket mouse	Arizona – SW Texas, N Mexico
P. lineatus	Lined pocket mouse	C Mexico
P. longimembris	Little pocket mouse	Utah – Baja California
P. nelsoni	Nelson's pocket mouse	N Mexico, SW Texas
P. parvus	Great Basin pocket mouse	Nevada – S Br. Colombia
P. penicillatus	Desert pocket mouse	S California – C Mexico
P. pernix	Sinaloan pocket mouse	NW Mexico
P. spinatus	Spiny pocket mouse	Baja California, S California
P. xanthonotus	Yellow-eared pocket mouse	C California

Microdipodops; kangaroo-mice; Nevada, etc; sand-steppes.

M. megacephalus	Dark kangaroo-mouse	Nevada, etc.
M. pallidus	Pale kangaroo-mouse	SW Nevada

Dipodomys; kangaroo-rats; SW USA – Mexico; desert, steppe.

D. agilis (paralius) (peninsularis)	Agile kangaroo-rat (Pacific kangaroo-rat)	C California – Baja California
D. californicus	Californian kangaroo-rat	N California, S Oregon
D. compactus	Gulf coast kangaroo-rat	S Texas, NE Mexico
D. deserti	Desert kangaroo-rat	Nevada – NW Mexico
D. elator	Texas kangaroo-rat	N Texas; *
D. elephantinus	Big-eared kangaroo-rat	WC California
D. gravipes	San Quintin kangaroo-rat	NW Baja California
D. heermanni	Heermann's kangaroo-rat	California; (*)
D. ingens	Giant kangaroo-rat	S California
D. insularis	San Jose kangaroo-rat	San Jose I, Baja California
D. margaritae (merriami)	Santa Margarita kangaroo-rat	Santa Margarita I, Baja California
D. merriami	Merriam's kangaroo-rat	Nevada – C Mexico
D. microps	Chisel-toothed kangaroo-rat	Nevada, etc.
D. nelsoni	Nelson's kangaroo-rat	NC Mexico; close to D. spectabilis
D. nitratoides	Fresno kangaroo-rat	C California
D. ordii	Ord's kangaroo-rat	W USA, NE Mexico
D. panamintinus	Panamint kangaroo-rat	California, SW Nevada; montane
D. phillipsi (ornatus)	Phillips' kangaroo-rat	C Mexico; (*)
D. spectabilis	Banner-tailed kangaroo-rat	New Mexico – C Mexico
D. stephensi	Stephens' kangaroo-rat	S California
D. venustus	Narrow-faced kangaroo-rat	WC California

Liomys; spiny pocket-mice; Mexico – Panama; arid scrub, steppe.

L. adspersus	Panamanian spiny pocket mouse	C Panama
L. irroratus	Mexican spiny pocket mouse	S Texas – S Mexico
L. pictus (annectens)	Painted spiny pocket mouse	W, S Mexico
L. salvini (crispus)	Salvin's spiny pocket mouse	S Mexico – Costa Rica
L. spectabilis	Jaliscan spiny pocket mouse	Jalisco, WC Mexico

Heteromys; forest spiny pocket mice; S Mexico – Ecuador; forest.

H. anomalus	Trinidad spiny pocket mouse	Colombia, Venezuela, Trinidad

H. australis	Southern spiny pocket mouse	E Panama – Ecuador
H. desmarestianus (*lepturus*) (*longicaudatus*) (*nigricaudatus*) (*temporalis*)	Desmarest's spiny pocket mouse	S Mexico – Colombia; ref. 9.9
H. gaumeri	Gaumer's spiny pocket mouse	Yucatan, Mexico
H. goldmani	Goldman's spiny pocket mouse	Guatemala, Chiapas
H. nelsoni	Nelson's spiny pocket mouse	Chiapas, S Mexico
H. oresterus	Mountain spiny pocket mouse	Costa Rica

Family Castoridae

Castor; beavers; 2 species; freshwater in deciduous forest.

C. canadensis	American beaver	N America, [Finland, Tierra del Fuego]
C. fiber (*albicus*)	Eurasian beaver	Europe – Altai

Family Anomaluridae

Scaly-tailed squirrels; 7 species; W, C Africa; forest; arboreal.

Anomalurus; (*Anomalurops*); large scaly-tailed squirrels; W, C Africa; (*).

A. beecrofti	Beecroft's flying squirrel	W Africa – Zaire
A. derbianus	Lord Derby's flying squirrel	Sierra Leone – Mozambique, Angola
A. pelii	Pel's flying squirrel	Sierra Leone – Ghana
A. pusillus	Little flying squirrel	Zaire, S Cameroun, Gabon

Idiurus; pygmy scaly-tailed squirrels; W, C Africa; (*).

I. macrotis	Long-eared flying squirrel	W Africa – Zaire
I. zenkeri	Zenker's flying squirrel	Cameroun, Zaire

Zenkerella

Z. insignis	Flightless scaly-tailed squirrel	Cameroun, Rio Muni, ? Gabon; forest

Family Pedetidae

Pedetes

P. capensis	Spring hare (Springhaas)	S Africa – S Kenya; steppe

Family Muridae

Mice, rats, voles, gerbils, hamsters, etc.; *c.* 1120 species; worldwide in all terrestrial (and freshwater) habitats; feed mainly on seeds, also grass, insects, etc.; classification very provisional – the groups recognized here as subfamilies are often treated as independent families, especially the Cricetidae and Spalacidae.

Subfamily Hesperomyinae

New-world mice and rats; *c.* 359 species; N, C, S America; desert – forest; classification of many Central and South American representatives very provisional.

Oryzomys; (*Melanomys, Nectomys, Nesoryzomys, Oecomys, Oligoryzomys, Sigmodontomys*); rice rats; S USA – Tierra del Fuego; grassland, marshes, scrub, (forest).

O. albigularis (*devius*) (*pirrensis*)	Tomes' rice rat	Borneo – Costa Rica; montane
O. alfari (*russulus*)	Alfaro's water rat	Honduras – Venezuela, Ecuador
O. alfaroi	Alfaro's rice rat	C Mexico – Ecuador
O. altissimus		Peru, Ecuador; montane
O. andinus		N Peru
O. aphrastus	Harris' rice rat	Costa Rica
O. arenalis		NE Peru
O. auriventer		Peru, Ecuador
O. balneator		Ecuador, E of Andes
O. bauri		Sante Fe I, Galapagos; (in *O. galapagoensis?*)
O. bicolor (*endersi*) (*phaeotis*) (*trabeatus*)		Bolivia, Amazon Basin – Panama; forest
O. bombycinus	Silky rice rat (Long-whiskered rice rat)	Nicaragua – N Ecuador
O. buccinatus		Paraguay, N Argentina
O. caliginosus	Dusky rice rat	Honduras – Ecuador
O. capito		Brazil, N Argentina – Venezuela

O. caudatus		Oaxaca, S Mexico
O. chacoensis		C Brazil – Argentina; ref. 9.81
O. chaparensis		E. Bolivia
O. cleberi		C Brazil; close to *O. bicolor*; ref. 9.10
O. concolor (*trinitatis*)		N Argentina, Bolivia – Costa Rica
O. couesi (*cozumelae*) (*palustris*) (*gatunensis*) (*? peninsulae*)		Panama – Texas; Jamaica
O. delicatus (*microtis*)		Colombia – Guyana, N Brazil
O. delticola		NE Argentina, Uruguay
O. dimidiatus	Thomas' water rat	Nicaragua
O. flavescens		S Brazil – N Argentina
O. fornesi		N Argentina
O. fulgens	Thomas' rice rat	C Mexico
O. fulvescens	Pygmy rice rat	C Mexico – Venezuela
O. galapagoensis †		San Cristobal I, Galapagos, probably extinct
O. gorgasi		NW Colombia; forest
O. hammondi		NW Ecuador
O. intectus		Colombia; montane
O. kelloggi		SE Brazil
O. lamia		SE Brazil
O. longicaudatus		Tierra del Fuego – Peru
O. macconnelli		Surinam – Ecuador
O. melanostoma		E Peru
O. melanotis	Black-eared rice rat	Mexico, El Salvador
O. minutus		Peru – Colombia; montane
O. munchiquensis		W Colombia
O. nelsoni	Nelson's rice rat	Maria Madre I, W Mexico
O. nigripes		E Brazil – Argentina
O. nitidus		Ecuador – N Argentina
O. palustris (*argentatus*)	Marsh rice rat	SE USA; marshes; ref. 9.11
O. polius		N Peru
O. ratticeps		SE Brazil, Paraguay, N Argentina
O. rivularis (*couesi*)		NW Ecuador

O. robustulus		E Ecuador
O. spodiurus		Ecuador
O. subflavus		E Brazil, Guianas
O. talamancae (*capito*)		Costa Rica – Ecuador; ref. 9.85
O. utiaritensis		C Brazil
O. victus †	St Vincent rice rat	St Vincent I, W Indies; extinct
O. xantheolus		Peru
O. yunganus		Bolivia, Peru
O. zunigae		Peru; coastal

Nesoryzomys; (*Oryzomys*); Galapagos mice; Galapagos Islands.

N. darwini †		Santa Cruz Is.; probably extinct
N. fernandinae		Fernandina Is.; ref. 9.12
N. indeffesus†		Santa Cruz Is.; probably extinct
N. narboroughi		Fernandina Is.
N. swarthi †		James Is.; probably extinct

Megalomys†; giant rice rats; Lesser Antilles; extinct.

M. desmarestii†	Antillean giant rice rat	Martinique, Lesser Antilles; extinct
M. luciae†	Santa Lucia giant rice rat	Santa Lucia, Lesser Antilles; extinct

Wiedomys; (*Oryzomys, Thomasomys*).

W. pyrrhorhinos	Red-nosed mouse	E Brazil; scrub

Neacomys; bristly mice (spiny rice rats); northern S America; forest.

N. guianae		Surinam – S Venezuela
N. spinosus		SW Brazil – Peru – Colombia
N. tenuipes (*pictus*) (*pusillus*)		E Panama – Ecuador, Venezuela

Scolomys

S. melanops	Ecuador spiny mouse	Ecuador

Nectomys; classification provisional – *N. squamipes* is probably composite.

N. parvipes		French Guiana
N. squamipes	South American water rat	S America N of Paraguay & S Brazil

Rhipidomys; American climbing mice; S America; forest.

R. latimanus (fulviventer) (venustus)		Colombia, Venezuela, Ecuador
R. leucodactylus		N Argentina – Venezuela
R. macconnelli		SE Venezuela, etc.
R. maculipes		E Brazil
R. mastacalis (venezuelae)		N Brazil, Venezuela, Guianas
R. scandens	Mt Pirri climbing mouse	E Panama
R. sclateri		Venezuela, Guyana, Trinidad

Thomasomys; (Wilfredomys); S America; forest.

T. aureus	Peru – Colombia; montane
T. baeops	W Ecuador
T. bombycinus	Colombia; montane
T. cinereiventer	Colombia, Ecuador; montane
T. cinereus	S Ecuador, N Peru
T. daphne	Bolivia, S Peru
T. dorsalis	E Brazil
T. gracilis	Ecuador, Peru
T. hylophilus	NE Colombia, W Venezuela
T. incanus	Peru
T. ischyurus	E Peru, E Ecuador
T. kalinowskii	Peru
T. ladewi	NW Bolivia; montane
T. laniger	W Venezuela, Colombia
T. monochromus (laniger)	NE Colombia
T. notatus	SE Peru; montane
T. oenax	S, E Brazil, Uruguay
T. oreas	Bolivia; montane
T. paramorum	Ecuador; montane
T. pictipes	NE Argentina; S Brazil
T. pyrrhonotus	S Ecuador, N Peru
T. rhoadsi	Ecuador
T. rosalinda	N Peru
T. taczanowskii	N Peru
T. vestitus	W Venezuela

Aepomys; (*Thomasomys*); N Andes.

A. fuscatus   Colombia
    (*lugens*)
A. lugens   Venezuela – Ecuador

Phaenomys
P. ferrugineus   Rio rice rat   E Brazil

Chilomys
C. instans   Colombian forest mouse   Ecuador – Venezuela;
                                                      montane forest

Tylomys; American climbing rats; S Mexico – Ecuador; forest.
T. bullaris   Chiapan climbing rat   Chiapas, S Mexico
T. fulviventer   Fulvous-bellied climbing   Panama
    (*mirae*)   rat
T. mirae   Colombia, N Ecuador
T. nudicaudus   Naked-tailed climbing rat   S Mexico – Nicaragua
    (*gymnurus*)
T. panamensis   Panama climbing rat   E Panama
T. tumbalensis   Tumbala climbing rat   Chiapas, S Mexico
T. watsoni   Watson's climbing rat   Costa Rica, Panama

Ototylomys
O. phyllotis   Big-eared climbing rat   S Mexico – Costa Rica

Nyctomys
N. sumichrasti   Sumichrast's vesper rat   S Mexico – Panama; forest

Otonyctomys
O. hatti   Yucatan vesper rat   Yucatan, Belize,
                                                      N Guatemala; forest

Rhagomys
R. rufescens   E Brazil; forest

Reithrodontomys; American harvest mice; N, C, (S) America; grassland, forest.
R. brevirostris   Short-nosed harvest mouse   Costa Rica, Nicaragua
R. burti   Sonoran harvest mouse   NW Mexico
R. chrysopsis   Volcano harvest mouse   C Mexico
R. creper   Chiriqui harvest mouse   Costa Rica, W Panama;
                                                      forest, terrestrial
R. darienensis   Darien harvest mouse   E Panama; forest, arboreal
R. fulvescens   Fulvous harvest mouse   S USA – Nicaragua
R. gracilis   Slender harvest mouse   S Mexico – Costa Rica
R. hirsutus   Hairy harvest mouse   WC Mexico

R. humulis	Eastern harvest mouse	SE USA
R. megalotis	Western harvest mouse	SW Canada – S Mexico
R. mexicanus	Mexican harvest mouse	C Mexico – Ecuador
R. microdon	Small-toothed harvest mouse	S Mexico – Guatemala
R. montanus	Plains harvest mouse	SC USA, NC Mexico
R. paradoxus		Nicaragua, Costa Rica
R. raviventris	Saltmarsh harvest mouse	San Francisco Bay; *
R. rodriguezi	Rodriguez's harvest mouse	Costa Rica
R. spectabilis	Cozumel harvest mouse	Cozumel I, Mexico
R. sumichrasti	Sumichrast's harvest mouse	C Mexico – Panama
R. tenuirostris	Narrow-nosed harvest mouse	Guatemala; montane

Peromyscus; deer mice, etc.; N, C America; forest – desert.

P. attwateri	Texas mouse	Texas, Oklahoma, etc.; forest, rocks
P. aztecus (*oaxacensis*)	Aztec mouse	C Mexico – Honduras; wet forest
P. boylii	Brush mouse	C, SW USA – Honduras
P. bullatus	Perote mouse	Veracruz, Mexico
P. californicus	California mouse	C, S California; scrub
P. caniceps	Burt's deer mouse	Monserrate I, Baja California
P. crinitus	Canyon mouse	Oregon – Colorado – NW Mexico; desert
P. dickeyi	Dickey's deer mouse	Tortuga I, Baja California
P. difficilis (*nasutus*)	Zacatecan deer mouse	Colorado – S Mexico; arid, rocky hills
P. eremicus (*collatus*)	Cactus mouse	Nevada – N Mexico
P. eva		Baja California
P. furvus (*latirostris*)	Blackish deer mouse	E Mexico; wet forest
P. gossypinus	Cotton mouse	SE USA; forest, swamps
P. grandis	Big deer mouse	Guatemala; montane forest
P. guardia	Angel Island mouse	Angel de la Guarda I, Baja California
P. guatemalensis (*altilaneus*)	Guatemalan deer mouse	S Mexico – Guatemala; montane forest
P. gymnotis (*allophylus*)		SE Mexico, Guatemala
P. hooperi	Hooper's mouse	Coahuila, N Mexico
P. interparietalis	San Lorenzo mouse	San Lorenzo Is, Baja California

P. leucopus	White-footed mouse	E, C USA – S Mexico; forest, scrub
P. madrensis	Tres Marias Island mouse	Tres Marias Is, W Mexico
P. maniculatus	Deer mouse	Labrador – Yukon – S Mexico
P. mayensis	Maya mouse	Guatemala; montane forest
P. megalops	Brown deer mouse	S Mexico; montane forest
P. mekisturus	Puebla deer mouse	Puebla, C Mexico; montane
P. melanocarpus	Zempoaltepec deer mouse	S Mexico; montane forest
P. melanophrys	Plateau mouse	C, S Mexico; steppe, desert
P. melanotis	Black-eared mouse	N, C Mexico; sibling of *P. maniculatus*
P. melanurus (*megalops*)	Black-tailed mouse	S Mexico
P. merriami	Merriam's mouse	N W Mexico, S Arizona; desert
P. mexicanus (*nudipes*)	Mexican deer mouse	E Mexico – W Panama; forest, scrub
P. ochraventer	El Carrizo deer mouse	NE Mexico; wet forest
P. pectoralis	White-ankled mouse	Texas, N, C Mexico; desert, scrub
P. pembertoni	Pemberton's deer mouse	San Pedro Nolasco I, N W Mexico
P. perfulvus	Marsh mouse	WC Mexico
P. polionotus	Oldfield mouse	SE USA
P. polius	Chihuahuan mouse	Chihuahua, Mexico
P. pseudocrinitus	False canyon mouse	Coronados I, Baja California
P. sejugis	Santa Cruz mouse	Santa Cruz & San Diego Is, Baja California
P. simulus		N W Mexico
P. sitkensis	Sitka mouse	Alexander & Queen Charlotte Is, Alaska/ W Canada
P. slevini	Slevin's mouse	Santa Catalina I, Baja California
P. spicilegus (*boylii*)		C Mexico
P. stephani	San Esteban Island mouse	San Esteban I, Baja California
P. stirtoni	Stirton's deer mouse	Honduras, El Salvador, Guatemala; montane

P. truei (*comanche*)	Pinyon mouse	SW USA – S Mexico; scrub, steppe
P. winkelmanni	Winkelmann's mouse	Michoacan, W Mexico; montane
P. yucatanicus	Yucatan deer mouse	Yucatan, Mexico; forest
P. zarhynchus	Chiapan deer mouse	SE Mexico; montane forest

Habromys; (*Peromyscus*).

H. chinanteco		Oaxaco, Mexico
H. lepturus (*ixtlani*)		Oaxaco, Mexico; montane forest
H. lophurus	Crested-tailed mouse	SE Mexico – El Salvador; montane
H. simulatus	Jico deer mouse	NW Mexico; forest

Isthmomys; (*Peromyscus*); Panama, etc.; montane.

I. flavidus	Yellow deer mouse	W Panama
I. pirrensis (*flavidus*)	Mount Pirri deer mouse	E Panama – NE Colombia

Megadontomys; (*Peromyscus*).

M. thomasi	Thomas's deer mouse	S Mexico; montane forest

Osgoodomys; (*Peromyscus*).

O. banderanus	Michoacan deer mouse	SW Mexico; dry forest

Podomys; (*Peromyscus*).

P. floridanus	Florida mouse	Florida

Ochrotomys

O. nuttalli	Golden mouse	SE USA; forest

Baiomys; American pygmy mice; C, (N) America; grassland.

B. musculus	Southern pygmy mouse	S Mexico – Nicaragua
B. taylori	Northern pygmy mouse	Texas, Arizona – C Mexico

Onychomys; grasshopper mice; W North America; scrub, steppe.

O. arenicola		New Mexico, N Mexico
O. leucogaster	Northern grasshopper mouse	SW Canada – N Mexico
O. torridus	Southern grasshopper mouse	SW USA – C Mexico

Akodon; (*Abrothrix, Chroeomys, Hypsimys, Thalpomys, Thaptomys*); South American field mice; S America; grassland, forest; classification very provisional.

A. aerosus (*urichi*)	Ecuador – Bolivia
A. affinis	Colombia
A. albiventer	Peru – N Argentina
A. andinus	C Chile – Peru; montane
A. azarae	NE Argentina – Bolivia – SE Brazil
A. boliviensis	S Peru – NW Argentina
A. budini	NW Argentina; montane
A. caenosus	NW Argentina, S Bolivia
A. cursor	Uruguay, Paraguay, SE Brazil
A. dolores	Sierra de Cordoba, C Argentina
A. hershkovitzi	Islands W of Tierra del Fuego; ref. 9.14
A. illutens	NW Argentina
A. iniscatus	SC Argentina
A. jelskii	NW Argentina – C Peru; montane
A. kempi	Islands in Parana Estuary
A. lanosus	Patagonia
A. llanoi	I de los Estados, Tierra del Fuego; ref. 9.15
A. longipilis	Chile, W, C Argentina
A. mansoensis	Rio Negro, C Argentina; ref. 9.16
A. markhami	Wellington I, S Chile
A. molinae	E Argentina
A. mollis	Ecuador – Bolivia
A. nigrita	E, S Brazil, N Argentina
A. olivaceus	Chile, W Argentina
A. orophilus	Peru
A. pacificus	W Bolivia; montane
A. puer	Bolivia, Peru; montane
A. reinhardti (*lasiotis*)	E Brazil
A. sanborni (*longipilis*)	Argentina, S Chile
A. serrensis	E, SE Brazil, N Argentina
A. surdus	SE Peru

A. urichi		Peru – Venezuela, Trinidad; montane
A. varius		W Argentina – Bolivia
A. xanthorhinus		Patagonia, Tierra del Fuego

Bolomys; (*Akodon*, *Cabreramys*); Peru – N Argentina; montane.

B. amoenus	SE Peru; montane
B. lactens	NW Argentina; montane
B. lasiurus (*lasiotis*) (*arviculoides*)	E Brazil, Paraguay
B. lenguarum (*tapirapoanus*)	Bolivia – Argentina
B. obscurus	Uruguay – C Argentina
B. temchuki	NE Argentina; ref. 9.79

Microxus; (*Akodon*).

M. bogotensis	Colombia, Venezuela
M. latebricola	Ecuador; montane
M. mimus	SE Peru; montane

Zygodontomys; cane mice; C, S America; savanna.

Z. borreroi	N Colombia
Z. brevicauda (*cherriei*) (*reigi*)	Costa Rica – Ecuador, Surinam, Trinidad

Podoxymys

P. roraimae	Roraima mouse	Guyana, etc.

Lenoxus

L. apicalis	Peruvian rat	Peru, Bolivia; montane forest

Oxymycterus; (*Akodon*); burrowing mice; S America S of Amazon; forest, cultivation.

O. akodontius	NW Argentina
O. angularis	E Brazil
O. delator	Paraguay
O. hispidus	NE Argentina – E Brazil
O. iheringi	S Brazil, N Argentina
O. inca	W Bolivia – C Peru
O. paramensis	SE Peru – N Argentina
O. roberti	E Brazil
O. rufus (*rutilans*)	C, E Brazil – NE Argentina

Juscelinomys
J. candango   Brasilia, Brazil

Blarinomys
B. breviceps   Brazilian shrew-mouse   SE Brazil; montane forest

Notiomys; long-clawed mice (mole-mice); southern S America; grassland, forest;
subterranean.
N. angustus   W Argentina
N. edwardsii   S Argentina
N. megalonyx   C Chile – Patagonia
    (delfini)

Geoxus; (Notiomys); ; predator; ref. 9.18.
G. valdivianus   Chile, S Argentina

Chelemys; (Notiomys); ref. 9.18.
C. macronyx   S Argentina, Chile

Kunsia; (Scapteromys).
K. fronto   SE Brazil, Paraguay,
                                                                N Argentina
K. tomentosus   S Brazil, Bolivia; savanna,
                                                                subterranean

Scapteromys
S. tumidus   Uruguay, NE Argentina,
                                                                etc.; swamps

Bibimys; (Akodon, Scapteromys); ref. 9.17.
B. chacoensis   N, C Argentina
B. labiosus   SE Brazil
B. torresi   Parana delta,
                                                                NE Argentina

Scotinomys; brown mice; central America; montane.
S. teguina   Alston's brown mouse   S Mexico – W Panama;
                                                                montane forest
S. xerampelinus   Chiriqui brown mouse   Costa Rica, W Panama;
                                                                montane forest, grass

Calomys; (Baiomys, Hesperomys); vesper mice; southern S America; grassland,
scrub.
C. callosus   S Brazil – Bolivia –
                                                                N Argentina; scrub

C. fecundus	Bolivia
C. hummelincki	Venezuela
C. laucha	C Argentina – SE Brazil; grassland, scrub
C. lepidus	Peru, Bolivia, etc.; montane
C. musculinus	Argentina
C. sorellus	Peru; montane

Eligmodontia; highland desert mice; Chile, etc.; arid scrub.

E. typus	S Peru – S Patagonia
(*hypogaeus*)	
(*puerulus*)	

Graomys; (*Phyllotis*).

G. domorum	N Argentina, Bolivia; montane
G. edithae	N Argentina
G. griseoflavus	S Argentina – Paraguay, Bolivia

Andalgalomys; (*Graomys*); ref. 9.77.

A. olrogi	NW Argentina; steppe
A. pearsoni	Paraguay, grassland

Pseudoryzomys; (*Oryzomys*).

P. simplex	E Brazil
P. wavrini	Paraguay, N Argentina, Bolivia; marsh

Phyllotis; (*Galenomys, Paralomys*); leaf-eared mice; S America.

P. amicus	N Peru
P. andium	Peru, Ecuador; forest
P. bonaeriensis	EC Argentina
(*darwini*)	
P. caprinus	NW Argentina, S Bolivia, E of Andes; scrub
P. darwini	Patagonia – W Peru
P. definitus	Peru, W of Andes; forest
P. garleppi	Bolivia, S Peru
P. gerbillus	NW Peru; desert
P. haggardi	Ecuador
P. magister	SW Peru, N Chile; forest
P. osilae	S Peru – N Argentina; montane grassland
P. wolffsohni	W Bolivia

Auliscomys; (*Phyllotis*); Andes; montane steppe.

A. boliviensis		S Peru – N Chile
A. micropus		Patagonia
A. pictus		Peru, W Bolivia
A. sublimus		NW Argentina – S Peru

Irenomys

I. tarsalis Chilean rat S Chile, S Argentina

Chinchillula

C. sahamae Chinchilla mouse NW Argentina – Peru

Punomys

P. lemminus Puna mouse Peru; high montane steppe

Neotomys

N. ebriosus Andean swamp rat Peru – NW Argentina;
montane

Reithrodon

R. physodes	Pampas gerbil	Uruguay – Tierra del
(*typicus*)	(Rabbit rat)	Fuego; grassland

Euneomys; (*Chelemyscus*); Patagonian chinchilla-mice; Argentina, Chile; scrub, forest.

E. chinchilloides		S Patagonia, Tierra del
(*petersoni*)		Fuego
E. mordax		W Argentina; (in
		E. chinchilloides?)

Holochilus; marsh rats; S America; marshes.

H. brasiliensis		E Brazil – C Argentina
H. chacarius		Paraguay, NE Argentina
H. magnus		SE Brazil – NE Argentina
H. sciureus		C Brazil – Colombia –
(*brasiliensis*)		Guianas

Sigmodon; cotton rats; S USA – Guyana; grassland, scrub.

S. alleni	Brown cotton rat	W, S Mexico
(*planifrons*)		
(*vulcani*)		
S. alstoni		Venezuela – Surinam
S. arizonae	Arizona cotton rat	Arizona, NW Mexico
S. fulviventer	Tawny-bellied cotton rat	SW Mexico – New Mexico
S. hispidus	Hispid cotton rat	S USA – Peru

S. leucotis (*alticola*)	White-eared cotton rat	C Mexico; montane
S. mascotensis		SW Mexico
S. ochrognathus	Yellow-nosed cotton rat	NC Mexico, S USA

Andinomys

A. edax	Andean mouse	Peru – NW Argentina, N Chile

Neotomodon

N. alstoni	Volcano mouse	C Mexico; montane

Neotoma; woodrats; N, C America; desert – forest, especially rocky.

N. albigula
    (*latifrons*)
    (*montezumae*)
	White-throated woodrat	SW USA, C Mexico

N. angustipalata	Tamaulipan woodrat	NE Mexico
N. anthonyi	Anthony's woodrat	Todos Santos I, Baja California
N. bryanti	Bryant's woodrat	Cerros I, Baja California
N. bunkeri	Bunker's woodrat	Coronados I, Baja California
N. chrysomelas	Nicaraguan woodrat	Nicaragua, Honduras; (in *N. mexicana*?)
N. cinerea	Bushy-tailed woodrat	S Yukon – Arizona
N. floridana	Eastern woodrat	SE USA; forest
N. fuscipes	Dusky-footed woodrat	Oregon – N Baja California
N. goldmani	Goldman's woodrat	NC Mexico

N. lepida
    (*devia*)
	Desert woodrat	SW USA, Baja California;
		    desert

N. martinensis	San Martin woodrat	San Martin I, Baja California
N. mexicana	Mexican woodrat	Colorado – Honduras
N. micropus	Southern plains woodrat	NE Mexico, W Texas, New Mexico, etc.
N. nelsoni	Nelson's woodrat	Veracruz, EC Mexico
N. palatina	Bolaños woodrat	Jalisco, W Mexico
N. phenax	Sonoran woodrat	NW Mexico; dry forest
N. stephensi	Stephen's woodrat	Arizona, W New Mexico
N. varia	Turner Island woodrat	Turner I, Gulf of California

Hodomys; (*Neotoma*).

H. alleni   Allen's woodrat   W, S Mexico; scrub

Nelsonia

N. neotomodon   Diminutive woodrat   WC Mexico; rocky forest

Xenomys

X. nelsoni   Magdalena rat   Jalisco, W Mexico; forest

Ichthyomys; fish-eating rats; NW South America; streams, swamps.

I. hydrobates   Ecuador, Colombia,
                                                        W Venezuela
I. pittieri   Venezuela
I. stolzmanni   Ecuador, Peru

Anotomys; (*Rheomys*).

A. leander   Ecuador fish-eating rat   Ecuador; montane streams
A. trichotis   Colombia, W Venezuela

Daptomys

D. oyapocki   French Guiana; ref. 9.19
D. peruviensis   Peruvian fish-eating rat   EC Peru; streams in
                                                        lowland forest
D. venezuelae   Venezuelan fish-eating rat   N Venezuela

Rheomys; water mice; C America; streams in forest.

R. hartmanni   Hartmann's water mouse   Costa Rica, W Panama
R. mexicanus   Mexican water mouse   Oaxaca, Mexico
R. raptor   Goldman's water mouse   E Panama
R. thomasi   Thomas' water mouse   El Salvador, Chiapas
R. underwoodi   Underwood's water mouse   Costa Rica, W Panama

Neusticomys

N. monticolus   Fish-eating mouse   Ecuador, Colombia;
                                                        montane

Subfamily Cricetinae

Hamsters etc.; *c.* 23 species; Europe, N Asia; grasslands, steppe; seed-eaters.

Calomyscus; mouse-like hamsters; species similar, sometimes all included in
C. bailwardi.

C. bailwardi   Iran, Afghanistan, etc.
C. baluchi   E Afghanistan,
                                                        W Pakistan

C. hotsoni		W Pakistan
C. mystax		S Turkmenia, etc.
C. urartensis		NW Iran, etc.

Phodopus

P. campbelli		Mongolia, etc. (in
(*sungorus*)		*P. sungorus*?)
P. roborovskii	Desert hamster	Mongolia – Shaanxi
P. sungorus	Striped hairy-footed hamster	E Kazakhstan, SW Siberia

Cricetus

C. cricetus	Common hamster	W Europe – Altai; grassland

Cricetulus; (*Allocricetulus, Tscherskia*); dwarf hamsters; SE Europe – China.

C. alticola	Ladak hamster	Ladak, Kashmir
C. barabensis	Striped hamster	S Siberia – Manchuria, Korea
C. curtatus	Mongolian hamster	Mongolia
C. eversmanni	Eversmann's hamster	N Kazakhstan
C. griseus (*barabensis*)		NE China
C. kamensis	Tibetan hamster	Tibetan plateau
C. longicaudatus	Lesser long-tailed hamster	NW China, Mongolia, etc.
C. migratorius	Grey hamster	SE Europe – Sinkiang
C. obscurus (*barabensis*)		Mongolia, NW China
C. pseudogriseus (*barabensis*)		N Mongolia, etc.
C. triton	Greater long-tailed hamster	NE China, Korea, Ussuri

Mesocricetus

M. auratus (*brandti*)	Golden hamster	Asia Minor, etc.
M. newtoni	Rumanian hamster	E Rumania, Bulgaria
M. raddei	Ciscaucasian hamster	Steppes N of Caucasus

Subfamily Spalacinae

Blind mole-rats; *c.* 8 species; E Mediterranean – S Russia; grassland, cultivation; subterranean; vegetarian.

Spalax

S. arenarius		S Ukraine

S. giganteus		N of Caspian Sea
S. graecus		N Rumania, Ukraine
S. microphthalmus		S Russia
S. polonicus		W Ukraine

Nannospalax; (*Spalax, Microspalax, Mesospalax*).

N. ehrenbergi (*leucodon*)		Syria – Libya
N. leucodon		Yugoslavia, Greece – SW Ukraine
N. nehringi (*leucodon*)		Asia Minor

Subfamily Myospalacinae

Zokors (eastern Asiatic mole-rats); *c.* 8 species; Altai – China; grassland, steppe.

Myospalax

M. aspalax (*myospalax*)		Transbaikalia, N Mongolia
M. baileyi (*fontanierii*)		Sichuan – Gansu, China
M. cansus (*fontanierii*)		Gansu – Shaanxi, China
M. fontanierii	Common Chinese zokor	Sichuan – Hebei
M. myospalax	Siberian zokor	Altai, W Siberia
M. psilurus (*myospalax*)		N China, Transbaikalia
M. rothschildi	Rothschild's zokor	Gansu, Hubei
M. smithii	Smith's zokor	Gansu

Subfamily Lophiomyinae

Lophiomys

L. imhausii	Crested rat	E Africa; montane forest

Subfamily Platacanthomyinae

Spiny dormice; 2 species; SE Asia; forest.

Platacanthomys

P. lasiurus	Malabar spiny dormouse	S India; forest, rocks

Typhlomys

T. cinereus	Chinese pygmy dormouse	SE China, N Indochina; montane forest

Subfamily Nesomyinae

Madagascan rats, etc.; *c*. 11 species; Madagascar, (S Africa); forest, grassland.

Macrotarsomys
M. bastardi   W Madagascar
M. ingens   NW Madagascar; forest

Nesomys
N. rufus   Madagascar

Brachytarsomys
B. albicauda   E Madagascar; forest

Eliurus
E. minor   E Madagascar; forest
E. myoxinus   Madagascar; forest

Gymnuromys
G. roberti   F. Madagascar

Hypogeomys
H. antimena   W Madagascar; sandy
                                                    coastal forest

Brachyuromys
B. betsileoensis   Madagascar
B. ramirohitra   E Madagascar

Mystromys
M. albicaudatus   White-tailed rat   S Africa

Subfamily Otomyinae

African swamp rats; *c*. 13 species; Africa S of Sahara; grassland, swamps.

Otomys, (*Myotomys*).
O. anchietae   Angola, S Tanzania
O. angoniensis   Angoni vlei rat   S Africa – Angola – Kenya
O. denti   NE Zambia – Uganda;
                                                    montane
O. irroratus   Vlei rat   S Africa – Zimbabwe
O. laminatus   Laminate vlei rat   S Africa
O. maximus   Large vlei rat   SW Zambia, etc.
O. saundersiae   Saunder's vlei rat   Cape Province, etc.
O. sloggetti   Sloggett's vlei rat   S Africa; montane

O. tropicalis		Cameroun – Kenya
O. typus		Zambia – Ethiopia; montane
O. unisulcatus	Bush vlei rat	Cape Province

Parotomys; Karroo rats.

P. brantsii	Brandt's whistling rat	Cape Province, etc.
P. littledalei	Littledale's whistling rat	Cape Province, Namibia

Subfamily Rhizomyinae

East African mole-rats, bamboo rats; *c.* 6 species; E Africa, SE Asia; montane forest, grassland, subterranean vegetarians.

Tachyoryctes

T. macrocephalus	Giant mole-rat	Ethiopia; montane grassland
T. splendens	East African mole-rat	E Africa; montane

Rhizomys; bamboo rats; SE Asia.

R. pruinosus	Hoary bamboo rat	Assam, S China – Malaya
R. sinensis		S China, N Burma
R. sumatrensis	Large bamboo rat	Indochina – Malaya, Sumatra

Cannomys

C. badius	Lesser bamboo rat (Bay bamboo rat)	Nepal – Thailand

Subfamily Microtinae (Arvicolinae).

Voles, lemmings; *c.* 121 species; N America, N Eurasia; tundra, grassland, scrub, open forest; herbivores.

Dicrostonyx; collared lemmings; tundra; classification very provisional.

D. groenlandicus (*torquatus*)	American arctic lemming	Greenland, N Canada, Alaska; possibly composite
D. hudsonius	Labrador collared lemming	Labrador
D. torquatus	Siberian arctic lemming (Collared lemming)	Siberia
D. vinogradovi	Wrangel lemming	Wrangel I, Siberia

Synaptomys; bog lemmings; N America.

S. borealis	Northern bog lemming	Canada, Alaska
S. cooperi	Southern bog lemming	NE USA, SE Canada

Myopus; (*Lemmus*).

M. schisticolor	Wood lemming	Scandinavia, Siberia; taiga taiga

Lemmus; brown lemmings; tundra.

L. amurensis	Amur lemming	Verkhoyansk Mts – Upper Amur
L. chrysogaster (*sibiricus*)		NE Siberia; (in *L. trimucronatus* ?)
L. lemmus	Norway lemming	Scandinavia
L. nigripes	Black-footed lemming	Pribilof Is
L. sibiricus	Siberian lemming	N Siberia W to White Sea
L. trimucronatus	Brown lemming	N, W Canada, Alaska

Clethrionomys; (*Evotomys, Eothenomys*); red-backed voles, bank voles; forest, scrub, tundra.

C. andersoni	Japanese red-backed vole	Honshu, Japan
C. californicus (*occidentalis*)	Western red-backed vole	Oregon – N California
C. centralis (*frater*)		Tien Shan Mts, etc.
C. gapperi	Gapper's red-backed vole	Canada (except NW), N USA, Rocky Mts, Appalachians; forest
C. glareolus	Bank vole	Europe – Altai; dec. forest
C. rufocanus (*rex*)	Grey red-backed vole	Scandinavia, Siberia, Hokkaido; tundra, montane
C. rutilus	Northern red-backed vole	N Eurasia, Arctic America; tundra, taiga

Eothenomys; (*Aschizomys*); oriental voles; E Asia.

E. chinensis		S China; montane
E. custos		S China; montane
E. eva		W China; montane
E. inez		C China
E. lemminus		NE Siberia; tundra
E. melanogaster	Père David's vole	S China, Taiwan
E. olitor		Yunnan, S China; montane
E. proditor		S China; montane
E. regulus		Korea, E Manchuria
E. shanseius		Shanxi, China
E. smithii		Japan, except Hokkaido

Alticola; mountain voles; C Asia.

A. macrotis	Large-eared vole	Altai, Sayan Mts
A. roylei	Royle's mountain vole	W Himalayas – Altai; montane
A. stoliczkanus (stracheyi)	Stoliczka's mountain vole	Himalayas – Altai
A. strelzowi	Flat-headed vole	Altai, E Kazakhstan

Hyperacrius

H. fertilis	True's vole	Kashmir, N Punjab
H. wynnei	Murree vole	N Pakistan; montane forest

Dinaromys; (Dolomys).

D. bogdanovi	Martino's snow vole	Yugoslavia; montane

Arvicola; water voles; N Eurasia; freshwater banks, grassland.

A. sapidus	Southwestern water vole	Iberia, SW France
A. terrestris (amphibius)	European water vole	Europe – E Siberia

Ondatra

O. zibethicus	Muskrat	N America, [N Eurasia]; freshwater banks

Neofiber

N. alleni	Round-tailed muskrat (Florida water rat)	Florida; freshwater banks, marshes

Phenacomys

P. intermedius	Heather vole	Canada, W USA; forest, tundra

Arborimus; (Phenacomys).

A. albipes	White-footed vole	W Oregon, NW California; coastal forest
A. longicaudus (silvicola)	Red tree vole	Oregon, NW California

Pitymys; (Blanfordimys, Microtus, Neodon); pine voles; N America, Eurasia; doubtfully distinct from Microtus.

P. afghanus	Afghan vole	Afghanistan, S Turkestan; montane
P. bavaricus	Bavarian pine vole	S Germany; montane

P. daghestanicus	Daghestan pine vole	E Caucasus
P. duodecimcostatus	Mediterranean pine vole	S, E Iberia – SE France
P. felteni (savii)		S Yugoslavia
P. gerbii (savii)		SW France, NE Spain
P. guatemalensis	Guatemalan vole	Guatemala; montane
P. juldaschi	Juniper vole	Tien Shan, Pamirs; montane
P. leucurus	Blyth's vole	Tibetan Plateau, Himalayas
P. liechtensteini	Liechtenstein's pine vole	NW Yugoslavia
P lusitanicus	Lusitanian pine vole	NW Iberia, SW France
P. majori	Major's pine vole	Caucasus, Asia Minor
P. multiplex	Alpine pine vole	European Alps
P. ochrogaster	Prairie vole	C, S USA
P. pinetorum	American pine vole	E USA; forest
P. quasiater	Jalapan pine vole	EC Mexico
P. savii	Savi's pine vole	Italy, S France
P. schelkovnikovi	Schelkovnikov's pine vole	Elburz, Talysh Mts, S of Caspian Sea
P. sikimensis	Sikkim vole	Himalayas – W China; montane
P. subterraneus	European pine vole	France – C Russia
P. tatricus	Tatra pine vole	Tatra Mts; montane
P. thomasi	Thomas' pine vole	S Balkans

Microtus; grass voles, meadow voles; N America, N Eurasia, (N Africa); grassland, open forest, tundra.

M. abbreviatus	Insular vole	St Matthew I, Hall I, Bering Sea; vicariant of *M. gregalis*
M. agrestis	Field vole	Europe – R Lena, E Siberia
M. arvalis	Common vole	Europe – R Yenesei; grassland
M. bedfordi	Duke of Bedford's vole	Gansu, China
M. brandtii	Brandt's vole	Mongolia, Transbaikalia
M. cabrerae	Cabrera's vole	Iberia
M. californicus	California vole	California
M. canicaudus	Grey-tailed vole	Oregon; close to *M. montanus*
M. chrotorrhinus	Rock vole	E Canada, NE USA

M. clarkei	Clarke's vole	Yunnan, N Burma; high montane
M. coronarius	Coronation Island vole	Coronation I, etc., Alaska
M. evoronensis		Lower Amur, E Siberia; ref. 9.20
M. fortis	Reed vole	China, SE Siberia
M. gregalis	Narrow-headed vole	Siberia, C Asia
M. gud		NE Asia Minor, Caucasus
M. guentheri	Gunther's vole	SE Europe – Israel, Libya
M. irani	Persian vole	Iran, etc.
M. kikuchii	Taiwan vole	Taiwan; montane
M. kirgisorum	Tien Shan vole	Tien Shan Mts
M. limnophilus (*oeconomus*)		W Mongolia, W China
M. longicaudus	Long-tailed vole	SW USA – Alaska
M. mandarinus	Mandarin vole	C China – SE Siberia
M. maximowiczii (*ungurensis*)		Upper Amur, E Siberia
M. mexicanus (*fulviventer*)	Mexican vole	Mexico, S USA; montane
M. middendorffii (*hyperboreus*)	Middendorff's vole	N Siberia; tundra
M. millicens	Szechwan vole	Sichuan, China
M. miurus	Singing vole	Alaska, Yukon
M. mongolicus	Mongolian vole	NE Mongolia, etc.
M. montanus	Montane vole	W USA; montane
M. montebelli	Japanese grass vole	Japan
M. mujanensis		Vitim Basin, E Siberia; ref. 9.78
M. nivalis	Snow vole	SW Europe – Iran; montane
M. oaxacensis	Tarabundi vole	Oaxaca, Mexico; montane forest
M. oeconomus	Root vole	N Eurasia, Alaska, etc.; tundra, taiga
M. oregoni	Creeping vole	W coast USA
M. pennsylvanicus (*breweri*)	Meadow vole	Canada, N USA; (*)
M. richardsoni	American water vole	NW USA, SW Canada; montane

M. rossiaemeridionalis (*epiroticus*) (*subarvalis*) (*arvalis*)		Finland, Greece – Urals, Caucasus; close to *M. arvalis*; ref. 9.21
M. roberti	Robert's vole	NE Asia Minor, W Caucasus; forest
M. sachalinensis	Sakhalin vole	Sakhalin I
M. socialis	Social vole	Kazakhstan – Palestine
M. townsendii	Townsend's vole	W coast USA, Vancouver I
M. transcaspicus	Transcaspian vole	SW Turkestan; steppe
M. umbrosus	Zempoaltepec vole	Oaxaca, Mexico; montane
M. xanthognathus	Yellow-cheeked vole	NW Canada, Alaska; forest, tundra

Lagurus; steppe lemmings; C Asia, W USA; steppe.

L. curtatus	Sagebrush vole	W USA; montane steppe
L. lagurus	Steppe lemming	Ukraine – Mongolia, Sinkiang

Eolagurus; (*Lagurus*); yellow steppe lemmings.

E. luteus		W Mongolia, N Sinkiang, († Kazakhstan)
E. przewalskii (*lutens*)		N Tibet – S Mongolia

Prometheomys

P. schaposchnikowi	Long-clawed mole-vole	Caucasus, NE Asia Minor

Ellobius; mole-voles; C Asia; steppe.

E. alaicus (*talpinus*)		Kirghizia
E. fuscocapillus	Southern mole-vole	SW Turkestan – Baluchistan
E. lutescens (*fuscocapillus*)		Caucasus – W Iran, etc.
E. talpinus	Northern mole-vole	Ukraine – Sinkiang

Subfamily Gerbillinae

˙Gerbils, jirds; *c.* 95 species; C, W Asia, Africa; steppe, desert; seed-eaters.

Gerbillus; (*Dipodillus, Monodia*); northern pygmy gerbils; N, E Africa, SW Asia; desert, steppe; classification very provisional; ref. 9.22.

G. acticola		Somalia
G. agag		Sudan, ? Chad
(? *dalloni*)		
G. allenbyi		Israel; coastal dunes
G. amoenus		Egypt – Mauritania
(*nanus*)		
G. andersoni	Anderson's gerbil	Israel – Tunisia
(? *bonhotei*)		
G. aquilus		W Pakistan – SE Iran
G. bottai		Sudan, Kenya
G. burtoni		Sudan
(*pyramidum*)		
G. campestris	Large North African gerbil	Morocco – Egypt, Sudan
(? *lowei*)		
G. cheesmani	Cheesman's gerbil	Arabia – SW Iran
G. cosensi		N Kenya
(*agag*)		
G. dasyurus	Wagner's gerbil	Arabia, Israel, Iraq
G. dunni		Ethiopia, Somalia
G. famulus	Black-tufted gerbil	SW Arabia
G. garamantis		Algeria
(*nanus*)		
G. gerbillus		Morocco – Israel
(*hirtipes*)		
G. gleadowi	Indian hairy-footed gerbil	NW India, Pakistan
G. grobbeni		NE Libya
(*nanus*)		
G. harwoodi		Kenya
G. henleyi	Pygmy gerbil	Algeria – Israel – Oman
G. hesperinus		Morocco
G. hoogstraali		Morocco
G. jamesi		Tunisia
G. latastei		Libya, Tunisia, ? Algeria
(*gerbillus*)		
(*pyramidum*)		
G. mackilligini		S Egypt, ? N Sudan
(*nanus*)		
G. maghrebi	Greater short-tailed gerbil	Morocco
G. mauritaniae		Mauritania

G. mesopotamiae	Mesopotamian gerbil	Iraq, SW Iran
G. muriculus		Sudan
G. nancillus		Sudan
G. nanus	Baluchistan gerbil	Morocco – Israel –
(? brockmani)		W Pakistan; ? Somalia
G. nigeriae		N Nigeria
(agag)		
G. occiduus		Morocco
G. perpallidus		N Egypt
G. poecilops	Large Aden gerbil	Arabia
G. principulus		Sudan
(nanus)		
G. pulvinatus		Ethiopia
(? bilensis)		
G. pusillus		Kenya, Ethiopia
(? diminutus)		
(? percivali)		
G. pyramidum	Greater Egyptian gerbil	Egypt, Sudan
(? dongolanus)		
(? floweri)		
G. riggenbachi		Rio de Oro, W Sahara
G. rosalinda		Sudan
G. ruberrimus		Somalia, Kenya
G. simoni	Lesser short-tailed gerbil	Algeria – Egypt
(kaiseri)		
(zakariai)		
G. somalicus		Somalia
(campestris)		
G. stigmonyx		Sudan
(campestris)		
G. syrticus		Libya
G. tarabuli		Libya
(pyramidum)		
G. vivax *		Libya
(nanus)		
G. watersi		Somalia, Sudan
(? juliani)		

Gerbillurus; (*Gerbillus*); southern pygmy gerbils; SW Africa; desert, subdesert.

G. paeba	Hairy-footed gerbil	SW Angola – Cape Province; desert, subdesert
G. setzeri	Setzer's hairy-footed gerbil	Namib Desert; gravel plains

G. tytonis		S Namib, SW Africa; desert
G. vallinus	Bushy-tailed hairy-footed gerbil	SW Angola – Orange R

Microdillus

M. peeli	Somalia pygmy gerbil	Somalia; semidesert

Tatera; rat-like gerbils; Africa, (SW Asia); steppe, savanna.

T. afra	Cape gerbil	SW Cape Province; steppe
T. boehmi	Boehm's gerbil	S Kenya – Zambezi, Angola
T. brantsi	Highveld gerbil	S Africa – Zambia
T. guineae (*robusta*)		W Africa; ref. 9.83
T. inclusa	Gorongoza gerbil	E Zimbabwe, Mozambique, E Tanzania
T. indica	Indian gerbil	W India – Asia Minor, Sri Lanka
T. leucogaster	Bushveld gerbil	S, C Africa; savanna
T. nigricauda	Black-tailed gerbil	Kenya
T. phillipsi (*robusta*)		Kenya, Ethiopia, Somalia; ref. 9.83
T. robusta	Fringe-tailed gerbil	Tanzania – Sudan; Burkina Faso
T. valida (*gambiana*)	Savanna gerbil	Senegal – Ethiopia – Angola; savanna

Taterillus; W, C, E Africa; savanna.

T. arenarius		Mauritania, Niger, Mali; sahel
T. congicus		Cameroun – Sudan
T. emini		Chad – Kenya, Ethiopia
T. gracilis		W Africa; guinea savanna
T. harringtoni		Sudan – NE Tanzania; dry savanna
T. lacustris		L Chad, etc.
T. pygargus		Senegal – Mali

Desmodillus

D. auricularis	Short-tailed gerbil (Cape short-eared gerbil)	S Africa

Desmodilliscus
D. *braueri* Pouched gerbil Sudan – Senegal

Pachyuromys
P. *duprasi* Fat-tailed gerbil N Sahara

Ammodillus
A. *imbellis* Somali gerbil Somalia, SE Ethiopia

Sekeetamys; (*Meriones*).
S. *calurus* Bushy-tailed jird E Egypt, S Israel –
   C Arabia

Meriones; jirds, gerbils; Sahara – China; steppe, desert.
M. *chengi* Turfan, Sinkiang
M. *crassus* Sahara – Afghanistan;
   desert
M. *hurrianae* Indian desert gerbil NW India – SE Iran
M. *libycus* Libyan jird Sahara – Sinkiang; desert
  (*caudatus*)
M. *meridianus* Mid-day gerbil Russian Turkestan – China
M. *persicus* Persian jird Asia Minor – Pakistan
M. *rex* King jird SW Arabia
M. *sacramenti* Buxton's jird S Israel
M. *shawii* Shaw's jird Morocco – Egypt
M. *tamariscinus* Tamarisk gerbil Lower Volga – W Gansu
M. *tristrami* Tristram's jird Sinai – Asia Minor,
   NW Iran
M. *unguiculatus* Mongolian gerbil Mongolia, Sinkiang –
   (Clawed jird) Manchuria
M. *vinogradovi* Vinogradov's jird E Asia Minor – N Iran
M. *zarudnyi* NE Iran, etc.

Brachiones
B. *przewalskii* Przewalski's gerbil Sinkiang – Gansu

Psammomys
P. *obesus* Fat sand rat Algeria – Arabia
P. *vexillaris* Algeria – Libya

Rhombomys
R. *opimus* Great gerbil Caspian Sea – Sinkiang,
   Pakistan

Subfamily Dendromurinae

African climbing mice, etc.; *c.* 22 species; Africa S of Sahara; forest – grassland.

Delanymys

D. brooksi		SW Uganda, etc.; montane swamps

Dendromus; African climbing mice; Africa S of Sahara; forest – grassland.

D. kahuziensis		E Zaire; montane
D. lovati		Ethiopia; montane
D. melanotis	Grey climbing mouse	S Africa – Ethiopia – Guinea; savanna
D. mesomelas	Brant's climbing mouse	S Africa – Ethiopia – Cameroun
D. messorius (*mystacalis*)		Nigeria – Zaire
D. mystacalis	Chestnut climbing mouse	S Africa – Ethiopia – Zaire
D. nyikae	Nyika climbing mouse	E Transvaal – Malawi – Angola

Dendroprionomys

D. rousseloti		Congo Republic; forest

Deomys

D. ferrugineus	Congo forest mouse	Zaire, Cameroun – Uganda; rain forest

Leimacomys

L. buettneri		Togo; forest

Malacothrix

M. typica	Large-eared mouse (Gerbil-mouse)	S Africa – S Angola

Megadendromus

M. nikolausi		E Ethiopia; montane; ref. 9.23

Petromyscus; rock mouse; SW Africa; rocky steppe.

P. collinus	Pygmy rock mouse	SW Cape Province – S Angola
P. monticularis	Brukkaros rock mouse	SW Africa

Steatomys; fat mice; Africa S of Sahara; savanna.

S. caurinus		Senegal – Nigeria

S. cuppedius		Senegal – Niger
S. jacksoni		Ghana, W Nigeria
S. krebsii	Krebs' fat mouse	S Africa – Angola, Zambia
S. parvus (minutus)	Tiny fat mouse	Senegal – Somalia – SW Africa; dry savanna
S. pratensis (caurinus)	Common fat mouse	Senegal – Sudan – Natal; savanna, dry forest

Prionomys

P. batesi	Dollman's tree mouse	Cameroun, Cent. African Rep.

Subfamily Cricetomyinae

African pouched rats; c. 5 species; Africa S of Sahara; savanna, forest.

Beamys; long-tailed pouched rats.

B. hindei (major)		S Kenya, Malawi; NE Zambia

Saccostomus; short-tailed pouched rats; savanna; ref. 9.24.

S. campestris	Pouched mouse	S Africa – Zambia
S. mcarnsi		Tanzania – Ethiopia, etc.

Cricetomys; giant pouched rats.

C. emini		Sierra Leone – L Tanganyika; rain forest
C. gambianus		Senegal – Sudan – S Africa; savanna

Subfamily Murinae

Old world mice and rats; c. 426 species; Eurasia, Africa, Australasia; forest (grassland); mainly seed-eaters.

Hapalomys; marmoset-rats; SE Asia; forest.

H. delacouri		Indochina, Hainan
H. longicaudatus		S Burma, Malaya

Vernaya; oriental climbing mice; S China, etc.; montane forest.

V. foramena	Sichuan climbing mouse	N Sichuan, China; ref. 9.25
V. fulva	Vernay's climbing mouse	S China, N Burma; montane forest

Tokudaia

T. osimensis	Ryukyu spiny rat	Ryukyu Is; forest; (*)

Vandeleuria

V. nolthenii		Sri Lanka; montane
V. oleracea	Palm mouse	India – Indochina, Sri Lanka; forest

Micromys

M. minutus (*danubialis*)	Harvest mouse	N Eurasia; grassland

Apodemus; (*Sylvaemus*); wood mice (long-tailed field mice); Palaearctic; forest, (grassland).

A. agrarius	Striped field mouse	C Europe – China
A. argenteus	Small Japanese field mouse	Japan
A. draco (*semotus*)		S China, etc., Taiwan
A. flavicollis	Yellow-necked mouse	Europe, Asia Minor
A. gurkha	Himalayan field mouse	Nepal; montane forest
A. latronum		SW China, N Burma
A. microps	Pygmy field mouse	E Europe, Asia Minor; grassland, scrub
A. mystacinus	Broad-toothed mouse	SE Europe, Asia Minor, etc., scrub, rocks
A. peninsulae	Korean field mouse	Manchuria, Korea, etc., Hokkaido
A. speciosus	Large Japanese field mouse	Japan
A. sylvaticus (*krkensis*)	Wood mouse	Europe – Altai, Himalayas; N Africa

Thamnomys; (*Grammomys*).

T. cometes	Mozambique woodland mouse	Kenya – Mozambique
T. dolichurus	Woodland mouse	Senegal – Ethiopia – S Africa; savanna, dry forest
T. rutilans		W Africa – Uganda, N Angola
T. venustus		E Zaire, etc.; montane forest

Carpomys; Luzon rats; Philippines; montane forest.

C. melanurus	Luzon, Philippines
C. phaeurus	Luzon, Philippines

Batomys; (*Mindanaomys*); Philippine forest rats; montane forest.

B. dentatus	Luzon, Philippines
B. granti	Luzon, Philippines
B. salomonseni	Mindanao, Philippines

Pithecheir

P. melanurus		Sumatra, Java; montane forest
P. parvus	Malayan tree rat (Monkey-footed rat)	Malaya

Hyomys

H. goliath	Rough-tailed giant rat (White-eared giant rat)	New Guinea

Conilurus

C. albipes	White-footed rabbit-rat	E, S Australia; probably extinct; *
C. penicillatus	Brush-tailed rabbit-rat	N, NW Australia, New Guinea; savanna

Zyzomys; (*Laomys*); Australian rock rats; N, C Australia; rocky outcrops in desert to savanna.

Z. argurus	Common Australian rock rat	W, N Australia; rocks
Z. pedunculatus	Macdonnell Range rock rat (Central rock rat)	C Australia; *
Z. woodwardi	Woodward's rock rat (Large rock rat)	NW, N Australia; rocks

Mesembriomys

M. gouldii	Black-footed tree rat	N Australia; savanna
M. macrurus	Golden-backed tree rat	NW, N Australia; savanna

Oenomys

O. hypoxanthus	Rufous-nosed rat	Sierra Leone – N Angola, E Africa; forest

Mylomys; (*Pelomys*).

M. dybowskyii		Ivory Coast – Kenya, Zaire; savanna

Dasymys

D. incomtus	African marsh rat (African water rat)	Senegal – Ethiopia – S Africa; marshes, river banks

Arvicanthis; African grass rats; Africa, Arabia; savanna; taxonomy very provisional.

A. abyssinicus	W Africa – Ethiopia – Zambia
A. blicki	Ethiopia; high montane
A. dembeensis (*testicularis*) (? *naso*)	Ethiopia – ? W Africa; ? SW Arabia
A. niloticus	Egypt
A. somalicus	Somalia, Ethiopia, Kenya

Hadromys

H. humei	Manipur bush rat	Assam, India; open montane forest

Golunda

G. ellioti	Indian bush rat	India, etc., Sri Lanka; marsh, cultivation

Pelomys; (*Desmomys*); groove-toothed swamp rats; Africa.

P. campanae		Zaire, W Angola
P. fallax	Creek rat	Kenya – Botswana
P. harringtoni		C Ethiopia; montane
P. hopkinsi		Rwanda, SW Uganda; marshes
P. isseli		Kome & Bussu Is, L Victoria
P. minor		Zaire, Angola, Zambia
P. rex		Ethiopia

Lemniscomys; striped grass rats; Africa; savanna.

L. barbarus		Senegal – Ethiopia, Tanzania, NW Africa; dry savanna, steppe
L. bellieri		Ivory Coast; savanna; ref. 9.26
L. griselda	Single-striped mouse	Angola
L. linulus (*griselda*)		Senegal – N Ivory Coast; close to *L. rosalia*
L. macculus		Uganda, etc.

L. mittendorfi (*striatus*)		Cameroun
L. rosalia (*griselda*)		Kenya – S Africa
L. roseveari		Zambia; ref. 9.27

L. striatus		Sierra Leone – Ethiopia –
		  Malawi – Angola;
		  savanna

Rhabdomys

R. pumilio	Four-striped grass mouse	S Africa – Uganda, Kenya; grassland

Hybomys; tropical Africa; forest.

H. lunaris (*univittatus*)		Ruwenzori Mts
H. univittatus		Guinea – Gabon

Typomys; (*Hybomys*); ref. 9.42.

T. trivirgatus		Guinea – Nigeria; forest

Millardia; Indian soft-furred rats; India, etc.; grassland, cultivation.

M. gleadowi	Sand-coloured rat	Pakistan, NW India
M. kathleenae		Burma
M. kondana		W India; ref. 9.28
M. meltada	Soft-furred field rat	India, Sri Lanka

Dacnomys

D. millardi	Millard's rat	Assam – Indochina; montane forest

Eropeplus

E. canus	Sulawesi soft-furred rat	Sulawesi

Stenocephalemys; Ethiopian narrow-headed rats; Ethiopia; high montane; should perhaps be included in *Praomys*.

S. albocaudata		Ethiopia
S. griseicauda		Ethiopia

Rattus; (*Nesoromys*); Old-world rats; SE Asia, Australia, some worldwide by association with man; mainly forest, scrub, cultivation; taxonomy still very provisional.

R. annandalei	Annandale's rat	S Thailand, Malaya, Sumatra

R. argentiventer	Ricefield rat	Indochina – Java, Borneo, Philippines, [Sulawesi, New Guinea]; cultivation, scrub
R. atchinus		Sumatra
R. baluensis	Summit rat	Sumatra, Borneo; montane forest
R. blangorum		Sumatra
R. bontanus		Sulawesi
R. callitrichus		Sulawesi
R. ceramicus	Ceram rat	Ceram, Indonesia; montane forest
R. culionensis		Culion, Philippines
R. dammermani		Sulawesi
R. doboensis		Aru Is, Indonesia
R. elaphinus		Talubu, Sulawesi
R. enganus		Enggano I (Sumatra)
R. everetti		Philippines
R. exulans (*bocourti*)	Polynesian rat	Bangladesh – New Guinea, New Zealand, Pacific islands; commensal
R. feliceus (*ruber*)		Ceram, Indonesia
R. foramineus		Sulawesi
R. fuscipes	Bush rat	E, S, SW Australia; coastal
R. giluwensis		C New Guinea; ref. 9.33
R. hamatus		Sulawesi
R. hoffmanni (*tatei*)		Sulawesi
R. hoogerwerfi		Sumatra
R. hoxaensis		N Vietnam
R. jobiensis		Islands in Geelvink Bay, New Guinea; ref. 9.33
R. koratensis (*sladeni*) (*sikkimensis*)	Sladen's rat	Nepal – Indochina; forest
R. leucopus	Mottle-tailed rat (Cape York rat) (White-footed spiny rat)	S New Guinea, N Queensland; lowland forest
R. losea (*exiguus*) (*sakeratensis*)	Lesser ricefield rat	SE China – Thailand, Taiwan; cultivation
R. lutreolus	Australian swamp rat	E, SE Australia, Tasmania
R. macleari		Christmas I, Indian Oc.

R. marmosurus (*facetus*)		Sulawesi
R. montanus		Sri Lanka; montane
R. morataiensis		Moluccas
R. mordax		E New Guinea; ref. 9.33
R. nativitatis		Christmas I, Indian Ocean
R. niobe	Moss-forest rat	New Guinea; montane

R. nitidus	Himalayan rat	Himalayas – Indochina,
		  Philippines, [Sulawesi,
		  New Guinea];
		  commensal

R. norvegicus	Brown rat
	  (Common rat)
	  (Norway rat)
		[Worldwide], original range
		  probably SE Asia;
		  commensal; includes
		  domestic rats

R. novaeguinae		NE New Guinea; ref. 9.33
R. osgoodi		S Vietnam; ref. 9.84
R. owiensis		Owi I, W New Guinea
R. palmarum		Nicobar Is

R. praetor
  (*ruber*)
	Variable spiny rat	Moluccas, New Guinea,
		  Solomon Is

R. pulliventer		Nicobar Is
R. punicans		Sulawesi
R. ranjiniae		S India
R. rattoides (*turkestanicus*)	Turkestan rat	Russian Turkestan – China

R. rattus	House rat
	  (Black rat)
	  (Ship rat)
		SE Asia, [worldwide in
		  tropics and warm
		  temperate zones];
		  commensal

R. richardsoni (*omichlodes*)	Richardson's rat	W New Guinea; montane
R. rogersi		S Andaman I
R. salocco		Sulawesi
R. simalurensis		Mentawei Is, Sumatra

R. sordidus
  (*colletti*)
	Australian dusky field rat
	  (Canefield rat)
		N, NE Australia, New
		  Guinea

R. steini (*leucopus*)		New Guinea; ref. 9.33
R. stoicus		Andaman Is
R. taerae		Sulawesi

R. tiomanicus
  (*jalorensis*)
	Malaysian field rat	Malaya – Borneo;
		  cultivation, secondary
		  forest

R. tunneyi	Tunney's rat (Pale field rat)	W, N, NE, C Australia; savanna, coastal sands
R. tyrannus		Philippines
R. verecundus	Slender rat	New Guinea
R. villosissimus	Long-haired rat	E, C Australia
R. xanthurus		Sulawesi

Maxomys; (*Rattus*); oriental spiny rats; SE Asia.

M. alticola	Mountain spiny rat	N Borneo; montane forest
M. baeodon	Small spiny rat	N Borneo
M. bartelsii		Java
M. dollmani		Sulawesi
M. hellwaldii		Sulawesi
M. hylomyoides		Sumatra
M. inas	Malayan mountain spiny rat	Malaya; montane forest
M. inflatus		Sumatra
M. moi		Indochina
M. musschenbroeki		Sulawesi
M. ochraceiventer	Chestnut-bellied spiny rat	Borneo; submontane forest
M. pagensis		Mentawai Is
M. panglima		Palawan etc.
M. rajah	Brown spiny rat (Rajah rat)	Malaya – Sumatra, Borneo; lowland forest
M. surifer	Red spiny rat	Indochina – Java, Borneo; lowland forest
M. whiteheadi	Whitehead's rat	Malaya, Sumatra, Borneo; forest

Sundamys; (*Rattus*); ref. 9.34.

S. infraluteus		Sumatra, Borneo
S. maxi		Java
S. muelleri		Malaya, Sumatra, Borneo, Palawan

Abditomys; (*Rattus*); ref. 9.35.

A. latidens		Luzon, Philippines

Palawanomys; ref. 9.34.

P. furvus		Palawan, Philippines

Niviventer; (*Rattus, Maxomys*); SE Asia; ref. 9.31.

N. andersoni		SW China
N. brahma		NE India, N Burma; montane forest

N. bukit	Malayan white-bellied rat	S Burma – Java; forest
N. confucianus	Chinese white-bellied rat	S Manchuria – N Indochina, Taiwan, Hainan
N. coxingi		N Burma, Taiwan
N. cremoriventer	Dark-tailed tree rat	Malaya – Java, Borneo; lowland forest
N. eha	Smoke-bellied rat	Nepal – SW China; montane
N. excelsior		SW China
N. fulvescens (*bukit*)	Chestnut rat	Himalayas – S China – Java; ref. 9.29
N. hinpoon		Thailand
N. langbianus		NE India – Thailand, Indochina; close to *N. cremoriventer*
N. lepturus		Java
N. niviventer	White-bellied rat	Himalayas
N. rapit	Long-tailed mountain rat	Thailand – Sumatra, Borneo
N. tenaster		Assam – Vietnam

Margaretamys; (*Rattus*); Sulawesi; forest; ref. 9.31.

M. beccarii		Sulawesi; lowland
M. elegans		Sulawesi; montane
M. parvus		Sulawesi; montane

Berylmys; (*Rattus*); ref. 9.34.

B. berdmorei	Small white-toothed rat (Grey rat)	S Burma, Thailand, Indochina
B. bowersi	Bower's rat	NW India – S China – Sumatra; forest
B. mackenziei	Kenneth's white-toothed rat	Assam – Thailand
B. manipulus	Manipur rat	Assam – Burma

Lenothrix; (*Rattus*).

L. canus	Grey tree rat	Malaya, Sumatra, Borneo; forest

Apomys; (*Rattus*); Philippines; ref. 9.37.

A. abrae		Luzon
A. datae		Luzon
A. hylocoetes (*petraeus*)		Mindanao

A. insignis		Mindanao
A. microdon		Catanduanas, etc.
A. musculus		Luzon, Mindoro
A. sacobianus		Luzon

Diplothrix; (*Rattus*).
D. legatus   Ryukyu Is

Leopoldamys; (*Rattus*); ref. 9.31.
L. edwardsi	Edwards' rat	Sikkim – S China – Sumatra
L. neilli	Neill's rat	S Thailand
L. sabanus	Long-tailed giant rat (Noisy rat)	Indochina – Java, Borneo; lowland forest
L. siporanus		Mentawei Is

Bunomys; (*Rattus*); ref. 9.32.
B. andrewsi (*adspersus*)	Sulawesi
B. chrysocomus	Sulawesi
B. fratrorum	Sulawesi
B. penitus	Sulawesi

Taeromys; (*Rattus*); ref. 9.32.
T. arcuatus	Sulawesi
T. celebensis	Sulawesi

Cremnomys; (*Rattus*); India, etc.
C. blanfordi	India, Sri Lanka
C. cutchicus	S, W India
C. elvira	SE India

Paruromys; (*Rattus*); ref. 9.32.
P. dominator   Sulawesi

Bullimus; (*Rattus*); Philippines; ref. 9.32.
B. bagobus (*bagopus*)	Mindanao
B. luzonicus	Luzon
B. rabori	Mindanao

Srilankamys; (*Rattus*); ref. 9.31.
S. ohiensis   Ohiya rat   Sri Lanka; montane

Kadarsanomys; (*Rattus*); ref. 9.36.
K. sodyi   W Java; montane forest

Anonymomys; ref. 9.31.
A. mindorensis — Mindoro, Philippines

Aethomys; (*Rattus*); Africa S of Sahara; savanna.

A. bocagei		Angola, Zaire
A. chrysophilus	Red veld rat	SW Africa – Natal – Kenya
A. granti	Grant's rock mouse	Karoo, S Africa
A. hindei		N Nigeria – NE Tanzania
A. kaiseri		E Zaire – Kenya – Malawi – Angola
A. namaquensis	Namaqua rock mouse	S Africa – S Zambia
A. nyikae (*dollmani*)	Nyika veld rat	NE Zambia, Malawi, E Zimbabwe
A. silindensis	Silinda rat	Mt Chirinda, Zimbabwe
A. thomasi		W C Angola

Thallomys; (*Aethomys, Rattus*).

T. paedulcus	Acacia rat (Black-tailed tree rat)	S Africa – Somalia; savanna, dry forest

Praomys; (*Hylomyscus, Mastomys, Myomyscus, Myomys*); African soft-furred rats; Africa (Arabia); mainly forest.

P. albipes	Ethiopia, etc.; forest, cultivation
P. angolensis	Angola, Zaire
P. coucha (*natalensis*)	S, ? E, ? W Africa; close to *P. natalensis*
P. daltoni (*butleri*)	Senegal – Sudan; savanna
P. delectorum	Kenya – Malawi; montane forest
P. derooi	Ghana – W Nigeria; savanna; ref. 9.38
P. erythroleucus (*natalensis*)	Morocco, Senegal – Sudan; steppe, savanna, cultivation
P. fumatus	E Africa, SW Arabia; steppe
P. hartwigi	Cameroun; montane forest
P. huberti (*erythroleucus*)	W Africa, Morocco, ? Somalia; guinea savanna; ref. 9.39
P. jacksoni	Nigeria – Sudan – Zambia; forest

P. morio		W C, Africa; forest
P. natalensis	Multimammate rat	S, ? E, ? W Africa; steppe, savanna, cultivation, commensal; close to *P. coucha*
P. pernanus		S Kenya, N Tanzania, Ruanda
P. rostratus		Ivory Coast, Liberia; ref. 9.40
P. ruppi		Ethiopia; montane; ref. 9.41
P. shortridgei	Shortridge's mouse	Namibia, Botswana
P. tullbergi		Senegal – Ghana; forest
P. verroxii (*verreauxi*)	Verreaux's mouse	Cape Province

Hylomyscus; (*Praomys*, *Rattus*); Africa S of Sahara; forest; classification very provisional.

H. alleni	Guinea – Gabon, etc.
H. baeri	Ivory Coast, Ghana
H. carillus (*aeta*)	Zaire, etc.
H. denniae	Uganda – Zambia
H. fumosus	Gabon, Cameroun, etc.
H. parvus	Zaire – Cameroun
H. stella	Nigeria – Kenya

Limnomys; (*Rattus*).

L. sibuanus (*mearnsi*)	Mindanao, Philippines; montane

Stochomys; (*Rattus*); ref. 9.42.

S. longicaudatus	Zaire – Nigeria; rain forest

Dephomys; (*Stochomys*); tropical Africa; forest; ref. 9.42.

D. defua	Guinea – Ivory Coast
D. eburnea (*defua*)	Ivory Coast

Tarsomys; (*Rattus*).

T. apoensis	Mindanao, Philippines

Tryphomys; (*Rattus*).

T. adustus	Mearns's Luzon rat	Luzon, Philippines; montane

Leporillus; *
L. apicalis	White-tipped stick-nest rat (Lesser stick-nest rat)	C Australia; dry forest, scrub; probably extinct
L. conditor	Greater stick-nest rat	Franklin I, S Australia (extinct mainland)

Leggadina; (*Pseudomys*); ref. 9.43.
L. forresti	Forrest's mouse	W, C Australia; arid zones
L. lakedownensis	Lakeland Downs mouse	NE Queensland

Pseudomys; (*Gyomys, Thetamys*); Australian mice; Australia; forest – desert.
P. albocinereus	Ashy-grey mouse	SW Australia; scrub
P. apodemoides (*albocinereus*)	Silky mouse	S Australia, Victoria
P. australis	Eastern mouse (Plains rat)	E, S, C Australia
P. chapmani	Pebble-mound mouse	NW Australia; ref. 9.44
P. delicatulus	Little native mouse (Delicate mouse)	NE, N, NW Australia, C New Guinea; steppe
P. desertor	Brown desert mouse	W, C Australia
P. fieldi	Alice Springs mouse	Alice Springs, C Australia; ? extinct; *
P. fumeus	Smokey mouse	Victoria; forest; *
P. glaucus (*albocinereus*)		S Queensland
P. gouldii	Gould's mouse	SW, S Australia; ? extinct
P. gracilicaudatus	Eastern chestnut mouse	C, S Queensland, New South Wales
P. hermannsburgensis	Sandy inland mouse	Australia; dry grassland, steppe
P. higginsi	Tasmanian mouse (Long-tailed mouse)	Tasmania; forest
P. nanus	Western chestnut mouse	W, N Australia; rocky hills
P. novaehollandiae	New Holland mouse	E New South Wales, Victoria, Tasmania
P. occidentalis	Western mouse	SW Australia; scrub
P. oralis	Hastings River mouse	SE Queensland; *
P. pilligaensis	Pilliga mouse	N New South Wales; ref. 9.45
P. praeconis	Shark Bay mouse	Shark Bay, W Australia; coastal dunes; *
P. shortridgei	Heath rat	Victoria, ? SW Australia; swamp, scrub; *

Melomys; (*Uromys*); mosaic-tailed rats; New Guinea, NE Australia; mainly montane forest.

M. aerosus		Ceram; montane
M. albidens	White-toothed melomys	W New Guinea; montane
M. arcium	Rossel Island melomys	Rossel I, E of New Guinea
M. burtoni (*littoralis*)	Grassland melomys	E, N Australia
M. capensis (*cervinipes*)	Cape York melomys	NE Queensland
M. cervinipes	Fawn-footed melomys	E, NE Australia; forest
M. fellowsi	Red-bellied melomys	NE New Guinea; montane
M. fraterculus		Ceram; montane
M. fulgens		Ceram, Talaud Is
M. hadrourus	Thornton Peak melomys	NE Queensland; ref. 9.46
M. leucogaster	White-bellied melomys	New Guinea
M. levipes	Long-nosed melomys	New Guinea
M. lorentzii	Long-footed melomys	New Guinea
M. lutillus	Little melomys	SE New Guinea, N Queensland
M. monktoni	Southern melomys	New Guinea
M. obiensis		Obi I, Halmahera
M. platyops	Lowland melomys	New Guinea
M. porculus		Guadalcanal, Solomon Is
M. rubex	Highland melomys	New Guinea
M. rubicola	Bramble Cay melomys	Bramble Cay (I), E New Guinea; (in *M. leucogaster*?)
M. rufescens	Rufescent melomys	New Guinea – Solomon Is

Pogonomelomys; (*Melomys*); New Guinea; forest; arboreal, terrestrial.

P. bruijnii	Lowland brush mouse	W, S New Guinea; lowland
P. mayeri	Shaw Mayer's brush mouse	W New Guinea; montane
P. ruemmleri	Rümmler's brush mouse	W New Guinea; montane
P. sevia	Highland brush mouse	NE New Guinea; montane

Solomys; (*Melomys*); naked-tailed rats; Solomon Islands; forest.

S. ponceleti		Bougainville I, Solomons
S. salebrosus		Bougainville I, Solomons
S. sapientis		Santa Ysabel I, Solomons

Uromys; giant naked-tailed rats; New Guinea, Solomon Islands, NE Australia; forest; arboreal.

U. anak (*neobritannicus*)	Black-tailed tree rat	C, NE New Guinea; New Britain; montane

U. caudimaculatus	Giant white-tailed rat (Mottle-tailed tree rat)	New Guinea, NE Queensland
U. imperator		Guadalcanal I, Solomons
U. rex		Guadalcanal I, Solomons
U. salamonis		Florida I, Solomons

Xenuromys
X. barbatus	Mimic tree rat	New Guinea

Malacomys; African swamp rats; W, C Africa; forest, swamp.
M. cansdalei		Ivory Coast – Ghana
M. edwardsi		Liberia – Nigeria
M. longipes		Cameroun – Uganda – NW Zambia
M. verschureni		E, NW Zaire

Haeromys; pygmy tree mice; Borneo, Sulawesi; forest.
H. margarettae	Ranee mouse	Borneo
H. minahassae		N Sulawesi
H. pusillus	Lesser ranee mouse	Borneo

Chiromyscus
C. chiropus	Fea's tree rat	Indochina, E Burma; forest

Diomys
D. crumpi	Crump's mouse	NE India, Nepal

Zelotomys
Z. hildegardeae		Angola – Malawi – Uganda – C Af. Rep.
Z. woosnami	Pale rat (Woosnam's desert rat)	NW Botswana, EC Namibia

Muriculus; (*Mus*).
M. imberbis		Ethiopia; montane

Mus; (*Coelomys, Gatimiya, Leggada, Leggadilla, Mycteromys, Nannomys*); SE Asia, Africa, [worldwide]; forest, grassland; ref. 9.49 (Asia); 9.48 (Europe).
M. abbotti	Abbott's mouse	Asia Minor, Greece, etc.
M. baoulei		Ivory Coast; ref. 9.47
M. booduga (*fulvidiventris*) (*lepidoides*)		C, S India, Sri Lanka
M. bufo		E Zaire
M. callewaerti		S Zaire, Angola

M. caroli	Ryukyu mouse	Ryukyu Is, Taiwan, Indochina – Sumatra, Java; cultivation
M. castaneus (*musculus*)	Oriental house mouse	India – Indonesia, Philippines
M. cervicolor	Fawn-coloured mouse	Nepal – Indochina, Sumatra, Java; cultivation, forest
M. cookii	Cook's mouse	Assam – Indochina
M. crociduroides		Sumatra; montane forest
M. domesticus (*musculus*)	House mouse	W, S Europe – Himalayas; [Americas, Australia]; commensal, outdoor forms scarce in range of other *Mus* spp.
M. dunni		India
M. famulus		S India; montane forest
M. fernandoni		Sri Lanka
M. goundae		N Cent. African Rep.
M. gratus		W Uganda
M. haussa		Senegal – Niger; steppe
M. hortulanus (*musculus*)	Steppe mouse	Austria – S Russia; cultivated land
M. indutus	Desert pygmy mouse	S Africa, Botswana
M. mahomet		Ethiopia
M. mattheyi		Senegal – Ghana; savanna
M. mayori		Sri Lanka
M. minutoides (*musculoides*)	Pygmy mouse	Africa S of Sahara; probably an aggregate of sibling species
M. musculus (*molossinus*)	E European house mouse	NE Europe, W Siberia, Japan
M. oubanguii		Cent. African Rep.; savanna
M. pahari	Gairdner's shrew-mouse	Sikkim – Indochina
M. phillipsi		India
M. platythrix		India
M. poschiavinus		Switzerland
M. proconodon (*pasha*)		Somalia, Ethiopia, Zaire
M. saxicola		India, Pakistan, Nepal
M. setulosus		Guinea – Gabon; Ethiopia
M. setzeri	Setzer's pygmy mouse	Botswana, Zambia; ref. 9.50

M. shortridgei	Shortridge's mouse	Burma – Indochina
M. sorella (*neavei*)		Uganda, Kenya – Zambia
M. spretus	Algerian mouse	SW Europe, NW Africa
M. tenellus		E Africa
M. terricolor		Nepal – Pakistan
M. triton	Grey-bellied pygmy mouse	E Zaire – Kenya
M. vulcani		Java; montane forest

Colomys: (*Nilopegamys*).
C. goslingi — Angola – Cameroun – Ethiopia; wet forest

Crunomys; ref. 9.51.

C. celebensis	C Sulawesi
C. fallax	Luzon, Philippines
C. melanius	Mindanao, Philippines
C. rabori	Leyte, Philippines

Archboldomys; ref. 9.51.
A. luzonensis — Luzon, Philippines

Macruromys

| *M. elegans* | Western small-toothed rat | W New Guinea; montane |
| *M. major* | Eastern small-toothed rat | NE New Guinea; montane |

Lorentzimys
L. nouhuysii — New Guinea jumping mouse — New Guinea; forest

Lophuromys; brush-furred rats; W, C, E Africa; forest, etc.

L. cinereus	E Zaire; montane
L. flavopunctatus	C, E Africa, Ethiopia
L. luteogaster	NE Zaire; forest
L. medicaudatus	C Africa; montane
L. melanonyx	Ethiopia; high montane
L. nudicaudus	Cameroun, etc.; forest-edge
L. rahmi	C Africa; montane forest
L. sikapusi	W, C Africa; forest savanna
L. woosnami	C Africa; montane forest

Notomys; Australian hopping mice; Australia; desert, steppe, (dry forest); *.
N. alexis — Spinifex hopping mouse — C, W Australia; desert, dry grassland

N. amplus	Short-tailed hopping mouse	C Australia; probably extinct
N. aquilo (*carpentarius*)	Northern hopping mouse	N Queensland, N Australia; coastal dunes
N. cervinus	Fawn hopping mouse	C Australia; desert
N. fuscus	Dusky hopping mouse	C Australia; desert, steppe
N. longicaudatus	Long-tailed hopping mouse	SW, C Australia; ? extinct
N. macrotis (*megalotis*)	Big-eared hopping mouse	SW Australia; ? extinct
N. mitchellii	Mitchell's hopping mouse	SE, S, SW Australia; dry forest – grassland

Mastacomys

M. fuscus	Broad-toothed rat	SE Australia, Tasmania; wet forest, marsh

Echiothrix

E. leucura	Sulawesi spiny rat (Celebes shrew-rat)	Sulawesi

Melasmothrix

M. naso	Lesser shrew-rat	Sulawesi

Tateomys; shrew rats; Sulawesi; ref. 9.51.

T. macrocercus		Sulawesi
T. rhinogradoides	Tate's rat	Sulawesi

Acomys; African spiny mice; Africa, SW Asia; desert – savanna; classification very provisional.

A. cahirinus (*cineraceus*) (*dimidiatus*)	Cairo spiny mouse	circum-Sahara, Ethiopia, Kenya, Israel – Pakistan; perhaps an aggregate of sibling species
A. cilicicus		Asia Minor; ref. 9.52
A. louisae (*subspinosus*)		Somalia; ref. 9.54
A. minous (*cahirinus*)		Crete
A. russatus	Golden spiny mouse	NE Egypt – Jordan, E Arabia
A. spinosissimus		Mozambique – Botswana
A. subspinosus		Transvaal – Somalia, Sudan

A. whitei		Oman; ref. 9.54
A. wilsoni		Kenya, S Sudan, S Ethiopia

Uranomys
U. ruddi	Rudd's mouse	Senegal – Kenya – Mozambique; savanna

Bandicota; (*Gunomys*); bandicoot-rats; SE Asia; scrub, cultivation, partly commensal.
B. bengalensis	Lesser bandicoot-rat	India – Burma, Sumatra, Java, Sri Lanka
B. indica	Greater bandicoot-rat	India – S China – Java, Sri Lanka, Taiwan
B. savilei		Burma – Indochina

Nesokia
N. indica	Short-tailed bandicoot-rat	Egypt – NW India, Sinkiang; steppe, cultivation

Erythronesokia; ref. 9.55.
E. bunni		S Iraq; marshes

Anisomys
A. imitator	New Guinea giant rat (Squirrel-toothed rat)	New Guinea; montane forest

Lenomys
L. meyeri (*longicaudus*)	Sulawesi giant rat	Sulawesi

Pogonomys; prehensile-tailed mice; New Guinea, etc.; forest; arboreal; ref. 9.56.
P. loriae (*fergussoniensis*) (*mollipilosus*)	Soft-haired tree mouse (Prehensile-tailed rat)	New Guinea, d'Entrecasteaux Is, N Queensland; montane
P. macrourus	Long-tailed tree mouse	New Guinea; lowland
P. sylvestris	Grey-bellied tree mouse	New Guinea; montane

Chiruromys; (*Pogonomys*); New Guinea etc.; forest; arboreal; ref. 9.56.
C. forbesi (*shawmayeri*)	Greater tree mouse (Forbes' tree mouse)	E New Guinea, etc.
C. lamia (*kagi*)	Broad-skulled tree mouse	SE New Guinea
C. vates	Lesser tree mouse	SE New Guinea; lowland

Chiropodomys; pencil-tailed tree mice; SE Asia; forest.

C. calamianensis	Palawan, etc., Philippines
C. gliroides (pusillus)	NE India – Java, Borneo
C. jingdongensis	Yunnan, China; montane; ref. 9.57.
C. karlkoopmani	Siberut & N Pagai Is, Mentawai Is
C. major	Borneo
C. muroides	Borneo

Mallomys

M. rothschildi	Smooth-tailed giant rat (Black-eared giant rat)	New Guinea; montane

Papagomys

P. armandvillei	Flores giant rat	Flores (Lesser Sunda Is)

Komodomys

K. rintjanus	Komodo rat	Komodo Is, Indonesia; ref. 9.58

Phloeomys; slender-tailed cloud rats; Luzon, Philippines; forest.

P. cumingi	Luzon, Philippines
P. pallidus	Luzon, Philippines; (in *P. cumingi* ?)

Crateromys; bushy-tailed cloud rats; Philippines; forest; ref. 9.59.

C. paulus	Ilin I, N Philippines
C. schadenbergi	Luzon, Philippines; montane forest

Subfamily Hydromyinae

Island water rats; *c.* 20 species; Philippines, New Guinea, Australia; forest, rivers, marshes; mainly predators; Philippine species should perhaps be in separate subfamily.

Chrotomys; Philippine striped rats; Philippines; ref. 9.60.

C. mindorensis	Mindoro striped rat	Mindoro, Luzon; lowland
C. whiteheadi	Luzon striped rat	Luzon, Philippines; montane

Celaenomys

C. silaceus	Luzon shrew-rat	Luzon, Philippines; montane forest

Crossomys
C. *moncktoni*   Earless water rat   E New Guinea; streams

Xeromys
X. *myoides*   False swamp rat   N, NE Australia; swamps; *

Hydromys; (*Baiyankamys*); beaver-rats; Australia, New Guinea, etc.; rivers; predators.

H. *chrysogaster*   Beaver-rat   Australia, Tasmania,
                    (Australian water rat)   New Guinea
H. *hubbemu*   Mountain water rat   W New Guinea
H. *hussoni*   W, C New Guinea;
                    ref. 9.61
II. *neobritannicus*   New Britain

Parahydromys
P. *asper*   Coarse-haired hydromyine   New Guinea; montane
                    forest

Neohydromys
N. *fuscus*   Short-tailed shrew-mouse   NE New Guinea; high
                    montane

Leptomys
L. *elegans*   Long-footed hydromyine   New Guinea

Paraleptomys; New Guinea; montane forest.
P. *rufilatus*   Red-sided hydromyine   N New Guinea
P. *wilhelmina*   Short-footed hydromyine   W New Guinea
                    (Short-haired
                    hydromyine)

Pseudohydromys; New Guinea; high montane forest.
P. *murinus*   Eastern shrew-mouse   NE New Guinea
P. *occidentalis*   Western shrew-mouse   W New Guinea

Microhydromys
M. *richardsoni*   Groove-toothed shrew-   W New Guinea
                    mouse

Mayermys
M. *ellermani*   One-toothed shrew-mouse   NE New Guinea; montane
                    forest

Rhynchomys; Luzon, Philippines; montane forest; ref. 9.62.

R. isarogensis	Isarog shrew-rat	Mt Isarog, Luzon
R. soricoides	Mount Data shrew-rat	Luzon

Family Gliridae (Muscardinidae)

Dormice; *c.* 16 species; Palaearctic, Africa; mainly forest, eating seeds and buds.

Glis; (*Myoxus*).

G. glis	Fat dormouse (Edible dormouse)	Europe, Asia Minor; forest

Muscardinus

M. avellanarius	Hazel dormouse	Europe, Asia Minor

Eliomys

E. quercinus (*melanurus*)	Garden dormouse	Europe, N Africa, Asia Minor – N Arabia; forest, scrub

Dryomys

D. laniger	Woolly dormouse	SW Asia Minor
D. nitedula	Forest dormouse	E Europe – Sinkiang

Glirurus

G. japonicus	Japanese dormouse	Japan, except Hokkaido

Chaetocauda

C. sichuanensis	Chinese dormouse	N Sichuan, China; ref. 9.82

Myomimus; mouse-tailed dormice.

M. personatus	N Iran, etc.
M. roachi	Bulgaria, etc., ?† Asia Minor, etc.
M. setzeri	Iran

Graphiurus; (*Claviglis*); African dormice; Africa S of Sahara; forest, savanna; classification very provisional.

G. crassicaudatus		Liberia – Cameroun
G. hueti		Guinea – Angola
G. murinus	Woodland dormouse	Africa S of Sahara
G. ocularis	Spectacled dormouse	SW Africa
G. parvus	Lesser savanna dormouse	Sierra Leone – Somalia – Zimbabwe
G. platyops	Rock dormouse	Zimbabwe – S Zaire – Namibia

Family Seleviniidae

One species.

Selevinia
S. betpakdalaensis   Desert dormouse   SE Kazakhstan; desert

Family Zapodidae

Jumping mice; *c.* 13 species; N Eurasia, N America; forest, grassland.

Sicista; birch mice; N Eurasia; forest, grassland.

S. betulina	Northern birch mouse	N, C Europe – E Siberia; forest
S. caucasica		Caucasus
S. concolor	Chinese birch mouse	Altai, Kashmir, W China; montane
S. kluchorica		W Caucasus; ref. 9.63
S. napaea	Altai birch mouse	N Altai Mts
S. pseudonapaea		S Altai Mts
S. subtilis	Southern birch mouse	E Europe – L Baikal; grassland
S. tianshanica	Tien Shan birch mouse	Tien Shan Mts

Zapus; American jumping mice; N America; wet grassland, forest-edge.

Z. hudsonius	Meadow jumping mouse	Canada, N, C USA
Z. princeps	Western jumping mouse	W North America
Z. trinotatus	Pacific jumping mouse	W coast USA

Eozapus
E. setchuanus   Sichuan jumping mouse   W, SW China; montane forest

Napaeozapus
N. insignis   Woodland jumping mouse   SE Canada, NE USA; forest

Family Dipodidae

Jerboas; *c.* 29 species; C Asia – NW Africa; desert, steppe.

Dipus
D. sagitta   Northern three-toed jerboa   S European Russia – N Iran – China

Paradipus
P. ctenodactylus   Comb-toed jerboa   SW Turkestan

Jaculus

J. blanfordi	Blanford's jerboa	E, S Iran, W Pakistan
J. jaculus (*deserti*)	Lesser Egyptian jerboa	Sahara, Arabia; desert
J. lichtensteini	Lichtenstein's jerboa	Caspian Sea – L Balkhash
J. orientalis	Greater Egyptian jerboa	Morocco – Israel; semidesert
J. turcmenicus	Turkmen jerboa	Turkmenistan, etc.

Stylodipus

S. telum	Thick-tailed three-toed jerboa	Ukraine – Mongolia

Allactaga

A. bobrinskii	Bobrinski's jerboa	Kizil-kum, Kara-kum Deserts, Turkestan
A. bullata		S, W Mongolia, etc.
A. elater	Small five-toed jerboa	E Asia Minor – Baluchistan; dry steppe
A. euphratica (*williamsi*)	Euphrates jerboa	Jordan – Afghanistan; dry steppe
A. firouzi		Isfahan, Iran; ref. 9.80
A. hotsoni	Hotson's jerboa	Persian Baluchistan
A. major	Great jerboa	Ukraine – Tien Shan
A. nataliae		Mongolia; ref. 9.64
A. severtzovi	Severtzov's jerboa	Turkestan
A. sibirica	Mongolian five-toed jerboa	R Ural – Manchuria; steppe
A. tetradactyla	Four-toed jerba	Libya, Egypt; coastal gravel plains

Alactagulus

A. pumilio (*pygmaeus*)		R Don – Inner Mongolia, N Iran

Pygeretmus; fat-tailed jerboas.

P. platyurus (*vinogradovi*)	Lesser fat-tailed jerboa	Kazakhstan
P. shitkovi	Greater fat-tailed jerboa	E Kazakhstan

Cardiocranius

C. paradoxus	Five-toed pygmy jerboa	Mongolia, etc.

Salpingotus; (*Salpingotulus*); three-toed pygmy jerboas; Mongolia – Baluchistan; ref. 9.65.

S. crassicauda	Thick-tailed pygmy jerboa	S Mongolia, E Turkestan; desert
S. heptneri	Heptner's pygmy jerboa	Uzbekistan, S of Aral Sea
S. kozlovi	Koslov's pygmy jerboa	S Mongolia; desert
S. michaelis	Baluchistan pygmy jerboa	NW Baluchistan
S. pallidus		Kazakhstan

Euchoreutes

E. naso	Long-eared jerboa	W Sinkiang – Inner Mongolia; desert

Family Hystricidae

Old-world porcupines; *c.* 11 species; Africa, S Asia; forest, savanna, steppe; ground vegetarians; ref. 9.66.

Hystrix; (*Acanthion, Thecurus*); short-tailed porcupines; Africa, S Asia, (S Europe); forest, savanna.

H. africaeaustralis		S, SE Africa; savanna
H. brachyura (*hodgsoni*)		Nepal – E China – Malaya, Sumatra, Borneo; forest
H. crassispinis		Borneo
H. cristata		W, E, N Africa, (S Europe); savanna, etc.
H. indica		SW Asia – India, Sri Lanka; steppe, scrub
H. javanica		Java – Flores, [Sulawesi ?]; forest
H. pumilis		Palawan, Philippines
H. sumatrae		Sumatra

Atherurus; brush-tailed porcupines; Africa, SE Asia; forest.

A. africanus	African brush-tailed porcupine	W, C, E Africa
A. macrourus	Asiatic brush-tailed porcupine	Assam – Malaya

Trichys

T. fasciculata (*lipura*)	Long-tailed porcupine	Malaya, Sumatra, Borneo; forest

Family Erethizontidae

New-world porcupines *c.* 10 species; N, C, S America; forest.

Erethizon

E. dorsatum	North American porcupine	Alaska – N Mexico; forest

Coendou; (*Sphiggurus*); tree porcupines; Mexico – Bolivia; forest.

C. bicolor (*rothschildi*)		Bolivia – Panama
C. insidiosus		E, C Brazil
C. mexicanus	Mexican porcupine	S Mexico – W Panama
C. prehensilis		N Argentina, Brazil – Venezuela
C. spinosus		E Brazil – N Argentina; (*)
C. vestitus (*pruinosus*)		Venezuela, Colombia
C. villosus		SE Brazil

Echinoprocta

E. rufescens	Upper Amazonian porcupine	Colombia; forest

Chaetomys

C. subspinosus	Thin-spined porcupine	E, N Brazil; scrub, *

Family Caviidae

Guinea pigs, etc.; *c.* 16 species; S America.

Cavia; guinea pigs, cavies; Colombia – N Argentina; delimitation of species very provisional.

C. anolaimae	Colombia
C. aperea	N Argentina – E Brazil
C. fulgida	E Brazil
C. guianae	Surinam – Venezuela
C. magna	S Brazil, Uruguay; ref. 9.67
C. nana	W Bolivia
C. tschudii	Ecuador – N Argentina; probable ancestor of domestic Guinea pig, *C. porcellus*

Kerodon

K. rupestris	Rock cavy	NE Brazil

Galea

G. *flavidens*	Brazil
G. *musteloides*	Peru – C Argentina
G. *spixii*	Brazil, E Bolivia
(*wellsi*)	

Microcavia

M. *australis*	Argentina
M. *niata*	Bolivia; montane
M. *shiptoni*	NW Argentina; montane

Dolichotis; (*Pediolagus*); maras (Patagonian cavies).

D. *patagonum*	Argentina
D. *salinicola*	NE Argentina, Paraguay, S Bolivia

Family Hydrochaeridae

One species.

Hydrochaerus

H. *hydrochaeris*	Capybara	E Argentina – Panama
(*isthmius*)		

Family Dinomyidae

One species.

Dinomys

D. *branickii*	Pacarana	Colombia Bolivia; forest

Family Dasyproctidae

Pacas, agoutis; *c.* 14 species; Mexico – S Brazil; forest, savanna.

Agouti; (*Cuniculus, Coelogenys, Stictomys*).

A. *paca*	S Mexico – Surinam – Paraguay, [Cuba]
A. *taczanowskii*	Venezuela – Ecuador; montane

Dasyprocta; agoutis; C, S America; forest, savanna.

D. *azarae*		C Brazil – N Argentina
D. *coibae*		Coiba I, Panama
D. *cristata*		Surinam
D. *fuliginosa*	Grey agouti	Colombia, upper Amazon, Surinam

D. guamara		Venezuela; swamp forest
D. kalinowskii		Peru
D. leporina (*albida*) (*aguti*)	Brazilian agouti	Venezuela – E Brazil, Lesser Antilles
D. mexicana	Mexican agouti	Veracruz, S Mexico, [Cuba]
D. punctata (*variegata*)		S Mexico – N Argentina, [Cuba]
D. ruatanica		Roatin I, Honduras

Myoprocta; acuchis; Amazon Basin; forest.

M. acouchy (*pratti*)		Amazon Basin
M. exilis		Amazon Basin

Family Chinchillidae

Viscachas, chinchillas; *c.* 6 species; southern S America.

Lagostomus

L. maximus	Plains viscacha	Argentina; grassland

Lagidium; mountain viscachas; Peru – Patagonia; montane.

L. peruanum		Peru
L. viscacia		Bolivia – N Chile, W Argentina
L. wolffsohni		S Chile, SW Argentina

Chinchilla; chinchillas; *.

C. brevicaudata		Peru – N Argentina
C. lanigera		Bolivia, Chile; montane rocks

Family Capromyidae

Hutias, etc.; 12 species; West Indies; forest, (freshwater margins).

Capromys; (*Geocapromys, Mesocapromys, Mysateles*); Cuban hutias; Cuba, Jamaica, etc.

C. angelcabrerae		Cay Ana Maria, S Cuba; ref. 9.68; *
C. arboricolus		Cuba; arboreal; ref. 9.69
C. auritus		Cay Fragoso, N Cuba; *
C. brownii	Brown's hutia	Jamaica, Swan I; *
C. garridoi		Canarreos Arch., Cuba; *

C. ingrahami	Bahaman hutia	Plana Keys, Bahamas, *
C. melanurus	Bushy-tailed hutia	E Cuba; arboreal; *
C. nanus	Dwarf hutia	Cuba; *
C. pilorides	Desmarest's hutia	Cuba
C. prehensilis	Prehensile-tailed hutia	Cuba; arboreal
C. sanfelipensis		Cay Juan Garcia, etc., W Cuba; *

Plagiodontia

P. aedium (hylaeum)	Hispaniolan hutia	Hispaniola; *

Family Myocastoridae

Myocastor

M. coypus	Coypu (Nutria)	Chile, Argentina – Bolivia, S Brazil; [Europe, N Asia, E Africa]

Family Octodontidae

Degus, etc.; c. 9 species; southern S America; steppe, mainly montane.

Octodon; degus; Peru, Chile.

O. bridgesi		Chile
O. degus		W Peru, Chile
O. lunatus		Chile, coastal hills

Octodontomys

O. gliroides	Mountain degu	Bolivia, N Chile, NW Argentina; montane steppe

Spalacopus

S. cyanus (tabanus)	Coruro	Chile; montane

Aconaemys

A. fuscus (porteri)	Chilean rock rat	SC Chile, W Argentina; forest; subterranean
A. sagei		Patagonia; ref. 9.18

Octomys; (Tympanoctomys).

O. barrerae		C Argentina
O. mimax	Viscacha-rat	W Argentina; montane

Family Ctenomyidae

Tuco-tucos; *c.* 36 species; Peru – Patagonia; subterranean; taxonomy very provisional.

Ctenomys

C. argentinus	Chaco, Argentina; ref. 9.71
C. australis	E Argentina; coastal
C. azarae	C Argentina
C. boliviensis	Bolivia, N Argentina
C. bonettoi	Chaco, Argentina; ref. 9.70
C. brasiliensis	E Brazil
C. colburni	SW Argentina
C. conoveri	Paraguay, N Argentina
C. dorsalis	Paraguay
C. emilianus	WC Argentina
C. flamarioni	S Brazil; ref. 9.72
C. frater	NW Argentina, S Bolivia
C. fulvus (**robustus**)	W Argentina, N Chile
C. knighti	W Argentina
C. latro	N Argentina
C. leucodon	Bolivia, Peru
C. lewisi	S Bolivia
C. magellanicus	Patagonia, Tierra del Fuego
C. maulinus	Chile
C. mendocinus (**haigi**)	C Argentina
C. minutus	S Brazil, Uruguay, N Argentina
C. nattereri	C Brazil
C. occultus	N Argentina
C. opimus	N Argentina, S Bolivia
C. perrensis	NE Argentina
C. peruanus	S Peru
C. pontifex	W Argentina
C. porteousi	C, E Argentina
C. saltarius	N Argentina
C. sericeus	SW Argentina
C. steinbachi	E Bolivia
C. talarum	E Argentina
C. torquatus	Uruguay, NE Argentina
C. tuconax	NW Argentina
C. tucumanus	. N Argentina

C. validus   C Argentina; ref. 9.73

Family Abrocomidae

Chinchilla-rats; 2 species; Peru – N Chile; grassland, scrub.

Abrocoma
A. bennetti   Chile
A. cinerea   S Peru – N Chile,
                                                NW Argentina

Family Echimyidae

American spiny rats; *c.* 45 species; C, S America; forest; taxonomy very
provisional; ref. 9.74.

Proechimys; terrestrial spiny rats; C, S America; mainly forest.
P. albispinus   E Brazil
P. amphichoricus   S Venezuela, etc.
P. bolivianus   Upper Amazon
P. brevicauda   E Peru, NW Brazil
P. canicollis   Colombia – Guyana
P. chrysaeolus   E Colombia
P. cuvieri   Guianas; ref. 9.75
P. decumanus   NW Peru, SW Ecuador;
                                                Pacific lowlands
P. dimidiatus   SE Brazil
P. goeldii   Amazonian Brazil
P. guairae   N Venezuela
P. gularis   Upper Amazon
P. guyannensis   Bolivia, Brazil – Colombia,
  (*warreni*)   Guianas
P. hoplomyoides   SE Venezuela, etc.
P. iheringi   E Brazil
P. longicaudatus   Brazil, etc.
P. magdalenae   Colombia
P. mincae   N Colombia
P. myosuros   E Brazil
P. oconnelli   E Colombia
P. oris   C Brazil
P. poliopus   NW Venezuela, etc.
P. quadruplicatus   Ecuador
P. semispinosus   Tomes' spiny rat   Honduras – Ecuador;
                                                Pacific coast only
P. setosus   E Brazil
P. simonsi   Peru – Colombia
  (*hendeei*)

P. steerei		Peru
P. trinitatis		Trinidad
P. urichi		N Venezuela

Hoplomys; doubtfully distinct from *Proechimys*.
H. gymnurus	Armoured rat	Honduras – Ecuador

Euryzygomatomys
E. spinosus	Guira	E, SE Brazil – N Argentina

Clyomys
C. bishopi	SE Brazil; ref. 9.76
C. laticeps	E, C Brazil, Paraguay

Carterodon
C. sulcidens	E Brazil

Thrichomys; (*Cercomys*).
T. apereoides (*cunicularis*)	NE Brazil – Paraguay

Mesomys
M. didelphoides	? Brazil
M. hispidus	Amazon Basin
M. obscurus	? Brazil
M. stimulax	N Brazil, Surinam

Lonchothrix
L. emiliae	Brazil S of Amazon

Isothrix; toros; N, C South America; river banks in forest.
I. bistriatus	Venezuela, Colombia, W Brazil
I. pictus	E Brazil
I. villosus	Peru

Diplomys; Panama – Ecuador; forest.
D. caniceps		Colombia, Ecuador
D. labilis (*darlingi*)	Gliding spiny rat	Panama
D. rufodorsalis		NE Colombia

Echimys; (*Makalata*); arboreal spiny rats; S America; forest.

E. armatus	Ecuador – Guianas, Trinidad
E. blainvillei	SE Brazil
E. braziliensis	E Brazil
E. chrysurus	NE Brazil, Surinam
E. dasythrix	SE, E Brazil
E. grandis	Amazon Basin
E. macrurus	Brazil S of Amazon
E. nigrispinus	E Brazil
E. saturnus	E Ecuador
E. semivillosus	Venezuela, Colombia
E. unicolor	? Brazil

Dactylomys; (*Lachnomys*); coro-coros; NW South America; forest; arboreal.

D. boliviensis	Bolivia, SE Peru
D. dactylinus	Colombia, Ecuador, Amazonian Brazil
D. peruanus	Peru; montane

Kannabateomys

K. amblyonyx	E Brazil – N Argentina

Thrinacodus

T. albicauda	Colombia
T. edax	W Venezuela

Family Thryonomyidae

Cane rats; *c.* 2 species; Africa S of Sahara; mainly marshes.

Thryonomys

T. gregorianus	Lesser cane rat	Cameroun – S Sudan – Zimbabwe
T. swinderianus	Greater cane rat	Africa S of Sahara

Family Petromyidae

One species.

Petromus

P. typicus	Dassie-rat (African rock rat)	Namibia, S Angola

Family Bathyergidae

African mole-rats; *c.* 9 species; Africa S of Sahara; desert, steppe, (forest); subterranean, feeding on roots, tubers, etc.

Georychus
G. capensis	Cape mole-rat	S Africa; grassland, cultivation

Cryptomys
C. hottentotus	Common mole-rat	S Africa – Zaire, Tanzania
C. mechowii		S Zaire, Zambia, etc.
C. ochraceocinereus	Ochre mole-rat	Ghana – N Uganda

Heliophobius
H. argenteocinereus	Silvery mole-rat	Zaire, Kenya – Zambezi
H. spalax	Thomas' silvery mole-rat	S Kenya

Bathyergus; dune mole-rats; S Africa; desert.
B. janetta	Namaqua dune mole-rat	Namaqualand, S Africa
B. suillus	Cape dune mole-rat	SW Cape Province, South Africa

Heterocephalus
H. glaber	Naked mole-rat	Ethiopia, Somalia, Kenya; steppe

Family Ctenodactylidae

Gundis; 5 species; Sahara, etc.; rocks in steppe and desert; herbivorous.

Ctenodactylus
C. gundi		Morocco – NW Libya
C. vali		S Morocco – Libya

Pectinator
P. spekei		Somalia, Ethiopia; steppe

Massoutiera
M. mzabi		C Sahara

Felovia
F. vae		Senegal, Mauritania, Mali

ORDER LAGOMORPHA

Lagomorphs; *c.* 59 species; Eurasia, Africa, N, C, (S) America; desert – forest; terrestrial herbivores.

Family Ochotonidae

Pikas; *c.* 17 species; N, C Asia, W N America; grassland, rocks, mainly montane; herbivores.

Ochotona

O. alpina		S Siberia, NE China, Hokkaido
O. collaris (*alpina*)	Collared pika	W Canada, Alaska
O. curzoniae	Black-lipped pika	Tibetan Plateau
O. daurica	Daurian pika	Altai – NE China
O. erythrotis	Chinese red pika	E Tibet, Qinghai, etc.
O. hyperborea (*alpina*)	Northern pika	NE Siberia
O. kamensis	Kam pika	Kam, W Sichuan
O. koslowi	Kozlov's pika	N Tibet
O. ladacensis	Ladak pika	S Sinkiang, Kashmir; high montane
O. pallasii	Pallas's pika	Altai – Tien Shan
O. princeps (*alpina*)	American pika	Alaska – New Mexico
O. pusilla	Steppe pika	Volga – Irtysh; grassland
O. roylei (*nepalensis*) (*macrotis*)	Royle's pika (Large-eared pika)	N Burma – Himalayas – Tien Shan; high montane
O. rufescens	Afghan pika	Afghanistan, Iran, etc.
O. rutila	Turkestan red pika	Tien Shan – Pamirs; montane
O. thibetana	Moupin pika	W China, Tibet
O. thomasi	Thomas' pika	Qinghai, China

Family Leporidae

Rabbits, hares; *c.* 42 species; Eurasia, Africa, N, C, (S) America; (desert), steppe, savanna, (forest), terrestrial herbivores; ref. 10.1

Pentalagus

P. furnessi	Ryukyu rabbit	Ryukyu Is.; *

Pronolagus; red rock rabbits, red hares; S, (E) Africa; savanna.

P. crassicaudatus	Greater red rock rabbit	E S Africa
P. randensis	Jameson's red rock rabbit	Zimbabwe – SW Africa
P. rupestris	Smith's red rock rabbit	S Africa – Kenya

Bunolagus; (*Lepus*).

B. monticularis	Riverine rabbit (Bushman hare)	Cape Province, S Africa

Romerolagus

R. diazi	Volcano rabbit (Zacatuche)	Mountains SE of Mexico City

Caprolagus

C. hispidus	Hispid hare	Nepal – Assam; *

Lepus; hares, jack rabbits; Eurasia, Africa, N America.

L. alleni	Antelope jack rabbit	NW Mexico, Arizona; steppe, desert
L. americanus	Snowshoe hare	Canada, N, W USA; open forest
L. brachyurus	Japanese hare	Japan (except Hokkaido)
L. californicus (*insularis*)	Black-tailed jack rabbit	C, SW USA, N Mexico; grassland, steppe
L. callotis (*mexicanus*)	White-sided jack rabbit	S Mexico – S New Mexico
L. capensis (*tolai*) (? *granatensis*)	Cape hare	Africa, S Europe – C China
L. comus (*oiostolus*)	Yunnan hare	Yunnan, China; ref. 10.4
L. europaeus (*capensis*) (? *castroviejoi*)	Brown hare	W Europe – Baikal; [SC Canada, NC USA, Argentina, Chile]
L. flavigularis	Tehuantepec jack rabbit	Oaxaca, S Mexico
L. habessinicus	Abyssinian hare	Ethiopia, Somalia
L. mandshuricus	Manchurian hare	Manchuria, etc.; forest
L. nigricollis	Indian hare	India, etc., Sri Lanka, [Java]
L. oiostolus	Woolly hare	Tibetan Plateau
L. peguensis (*siamensis*)	Burmese hare	Burma – Indochina, Hainan
L. saxatilis (*whytei*) (*crawshayi*)	Scrub hare	S Africa – Sudan – Senegal; savanna

L. sinensis	Chinese hare	S China, Korea, Taiwan
L. timidus	Arctic hare	N Eurasia, arctic
(*arcticus*)	(Mountain hare)	N America, Greenland; tundra, open forest
L. townsendi	White-tailed jack rabbit	C, W USA; grassland, steppe
L. yarkandensis	Yarkand hare	SW Sinkiang; steppe

Poelagus

P. marjorita	Central African hare	Angola – Sudan; savanna

Sylvilagus; cottontails, American rabbits; N, C, (S) America.

S. aquaticus	Swamp rabbit	Texas – Georgia
S. audubonii	Desert cottontail	C, SW USA – C Mexico; desert, steppe
S. bachmani	Brush rabbit	W coast USA, Baja California; scrub
S. brasiliensis	Forest rabbit	NE Mexico – N Argentina; forest, scrub
S. cunicularius	Mexican cottontail	S Mexico
S. dicei		Costa Rica, W Panama
S. floridanus	Eastern cottontail	E, C USA – Costa Rica
S. graysoni	Tres Marias cottontail	Tres Marias Is, Mexico
S. insonus	Omilteme cottontail	Guerrero, Mexico; montane
S. nuttalli	Nuttall's cottontail	W USA; scrub, forest
S. palustris	Marsh rabbit	Florida – SE Virginia
S. transitionalis	New England cottontail	Appalachian Mts – S Maine

Oryctolagus

O. cuniculus	European rabbit	Iberia, NW Africa, [Europe, Australia, New Zealand, Chile]; grassland, open forest, cultivation; includes domestic rabbits

Brachylagus; (*Sylvilagus*); ref. 10.2.

B. idahoensis	Pygmy rabbit	Montana, Oregon, Idaho, etc.; steppe, scrub

Nesolagus

N. netscheri	Sumatran rabbit	Sumatra; montane forest; *

ORDER MACROSCELIDEA

One family, sometimes included in order Insectivora.

Family Macroscelididae

Elephant-shrews; *c.* 15 species; Africa (except west); subdesert, steppe, savanna, forest; terrestrial; insectivores; ref. 10.3.

Macroscelides

M. proboscideus	Short-eared elephant-shrew	SW, S Africa; steppe

Elephantulus; (*Nasilio*); small elephant-shrews; Africa; savanna – subdesert.

E. brachyrhynchus	Short-snouted elephant-shrew	E, S Africa
E. edwardii	Cape elephant-shrew	S Africa
E. fuscipes	Uganda elephant-shrew	Uganda, etc.
E. fuscus	Zambesi elephant-shrew (Peters' short-snouted elephant-shrew)	Lower Zambezi
E. intufi	Bushveld elephant-shrew	Namibia, etc.
E. myurus	Transvaal elephant-shrew	Zimbabwe – Cape Prov.
E. revoilii	Somalia elephant-shrew	Somalia
E. rozeti	North-African elephant-shrew	NW Africa
E. rufescens	Rufous elephant-shrew	E Africa
E. rupestris	Rock elephant-shrew	SW, S Africa

Petrodromus

P. tetradactylus (*sultani*) (*tordayi*)	Four-toed elephant-shrew	E, SE, C Africa; savanna

Rhynchocyon; (*Rhinonax*); forest elephant-shrews; E, C Africa; forest.

R. chrysopygus	Yellow-rumped elephant-shrew	Kenya coast
R. cirnei (*stuhlmanni*)	Chequered elephant-shrew	E, C Africa
R. petersi	Black and rufous elephant-shrew	SE Kenya, E Tanzania coast

Bibliographies

These bibliographies are arranged in three groups: general works on the diversity and classification of mammals; regional works, which are usually the best guides to identification; and works on particular groups of mammals. The first two are arranged in chronological order beginning with the most recent. The numbered taxonomic sources are arranged in a sequence corresponding to the main text and include all items referred to by number in the text. These include the original description of all species newly described in the last five years, along with the sources for classifications that do not follow those found in the general and regional works listed below.

General Works

Anderson, S. & Jones, J.K. (Eds.) 1984. *Orders and families of recent mammals of the world.* New York etc.: Wiley, 686 pp.

Macdonald, D. (Ed.) 1984. *The encyclopaedia of mammals*, 2 vols. London: Allen & Unwin, 895 + xxxii pp.

Burton, J.A. 1984. Bibliography of Red Data Books. *Oryx*, **18**: 61–64.

Nowak, R.M. & Paradiso, J.L. 1983. *Walker's mammals of the world*, 4th ed. Baltimore & London: Johns Hopkins University Press, 1362 pp. (2 vols.)

Honacki, J.H., Kinman, K.E. & Koeppl, J.W. (Eds.) 1982. *Mammal species of the world.* Lawrence, Kansas: Allen Press, ix + 694 pp.

Thornback, J. & Jenkins, M. 1982. *The IUCN mammal red data book. Part 1. Threatened mammalian taxa of the Americas and the Australian zoogeographic region (excluding Cetacea).* Gland, Switzerland: IUCN, 516 pp.

Hickman, G.C. 1981. National mammal guides: a review of references to Recent faunas. *Mammal Review*, **11**: 53–85.

McKenna, M.C. 1975. Towards a phylogenetic classification of the Mammalia. *In* Luckett, W.P. & Szalay, F.S. (Eds.) *Phylogeny of the Primates.* New York: Plenum Publishing Co.: 21–46. (A provisional classification of fossil and recent mammals to ordinal level.)

Anderson, S. (Ed.) 1969–. *Mammalian species.* American Society of Mammalogists. (A loose-leaf part-work, each part dealing with a single species in 2–8 pages; 239 parts issued by July 1985.)

Simpson, G.G. 1945. The principles of classification and a classification of the mammals. *Bulletin of the American Museum of Natural History*, **85**: i–xvi, 1–350. (Lists all recent and fossil genera.)

Geographical sources

In the following lists priority has been given to selecting reference works that are recent, authoritative with respect to taxonomy, comprehensive in coverage of species, useful for identification and that cover a wide area and include distribution maps, although few meet all of these criteria.

Palaearctic Region (Europe, N. Asia, N. Africa)

Corbet, G.B. 1984. *The mammals of the Palaearctic Region: a taxonomic review: supplement.* London: British Museum (Natural History), 45 pp.

Niethammer, J. & Krapp, F. 1978, 1982. *Handbuch der Säugetiere Europas.* Vols. 1 and 2 (1): Rodentia. Wiesbaden: Akademische Verlagsgesellschaft, 476 + 649 pp.

Pucek, Z. (Ed.) 1981. *Keys to the vertebrates of Poland: mammals.* Warsaw: translated and published for Smithsonian Institution and National Science Foundation, Washington DC by Polish Scientific Publishers, 367 pp.

Gromov, I.M. & Baranova, G. 1981. [*Catalogue of mammals of the USSR: Pliocene to the present day*]. Leningrad: Akademia Nauk SSSR. 455 pp.

Corbet, G.B. & Ovenden, D. 1980. *The mammals of Britain and Europe.* London: Collins, 253 pp.

Sokolov, V.E. & Orlov, V.N. 1980. [*Key to the mammals of the Mongolian People's Republic.*] Moscow, 352 pp. (In Russian).

Corbet, G.B. 1978. *The mammals of the Palaearctic Region: a taxonomic review.* London: British Museum (Natural History), 314 pp. (Includes keys and distribution maps for all species.)

Corbet, G.B. & Southern, H.N. (Eds.) 1977. *The handbook of British mammals,* 2nd ed. Oxford etc: Blackwell, 520 pp.

Saint Girons, M.-C. 1973. *Les mammifères de France et du Benelux.* Paris: Doin, 481 pp.

Bobrinski, N.A., Kuznetsov, B.A. & Kuzyakin, A.P. 1965 *Opredelitel' mlekopitayushchikh SSSR.* [Key to mammals of the USSR]. Moscow. (Coloured illustrations, text in Russian.)

Harrison, D.L. 1964–72. *The mammals of Arabia.* London: Ernest Benn, 3 vols.

Imaizumi, Y. 1960. *Coloured illustrations of the mammals of Japan.* Osaka: Hoikusha Publishing Co., 196 pp. (Text in Japanese.)

Shou, Z.-H. 1964. [*Handbook of economic animals of China. Mammals*]. Peking, 554 pp, 72 pls.

Australasian Region

Strahan, R. (Ed.) 1983. *Complete book of Australian mammals.* London etc: Angus & Robertson, 530 pp.

Ziegler, A.C. 1982. An ecological check-list of New Guinea recent mammals, pp. 863–894. *In* Gressitt, J.L. (Ed.) *Biogeography and ecology of New Guinea,* vol. 2. Monographiae Biologicae, vol. 42. The Hague: W. Junk, 983 pp.

Tyler, M.J. 1979. *The status of endangered Australian wildlife.* Adelaide: Royal Zoological Society of South Australia, 210 pp.

Oriental Region (SE Asia)

Medway, Lord. 1983. *The wild mammals of Malaya* .... 2nd ed. (revised). Kuala Lumpur etc: Oxford University Press.

Phillips, W.W.A. 1980–84. *Manual of the mammals of Sri Lanka,* 2nd ed. In 3 parts. Sri Lanka: Wildlife and Nature Protection Society, 390 pp.

Medway, Lord 1977. *Mammals of Borneo: field keys and an annotated checklist.* Kuala Lumpur: Malaysian Branch Royal Asiatic Society (monograph no.7) xii + 172 pp.

Lekagul, B. & McNeely, J.A. 1977. *Mammals of Thailand.* Bangkok. (All species, with illustrations, keys and maps.)

Roberts, T.J. 1977. *The mammals of Pakistan.* London: Ernest Benn, 361 pp. (Includes keys, illustrations and maps.)

Eisenberg, J.F. & McKay, G.M. 1970. An annotated checklist of the mammals of Ceylon with keys to the species. *Ceylon Journal of Science (Biol.),* **8**: 69–99.

Van Peenen, P.F.D., Ryan, P.F. & Light, R.H. 1969. *Preliminary identification manual for mammals of South Vietnam.* Washington: Smithsonian Institution, 310 pp.

Ellerman, J.R. & Morrison-Scott, T.C.S. 1951. *Checklist of Palaearctic and Indian mammals 1758–1946.* London: British Museum (Natural History), 810 pp. (Covers Oriental Region north of 10°N in Malaya.)

Africa

Smithers, R.H.N. 1983. *The mammals of the southern African subregion.* Pretoria: University of Pretoria, xxii + 736 pp.

Kingdon, J. 1971–82. *East African mammals: an atlas of evolution in Africa*, 7 vols. London etc: Academic Press.

Haltenorth, T. & Diller, H. 1980. *A field guide to the mammals of Africa including Madagascar*. London: Collins, 400 pp.

Osborn, D.J. & Helmy, I. The contemporary land mammals of Egypt (including Sinai). *Fieldiana: Zoology*, N.S. **5**: xix + 579 pp.

Swanepoel, P., Smithers, R.H.N. & Rautenbach, I.L. 1980. A checklist and numbering system of the extant mammals of the southern African subregion. *Annals of the Transvaal Museum*, **32**: 155–196.

Ansell, W.F.H. 1978. *The mammals of Zambia*. Chilanga: Department of National Parks & Wildlife Service, 126 pp. + 204 maps.

Largen, M.J., Kock, D. & Yalden, D.W. 1974. Catalogue of the mammals of Ethiopia. 1. Chiroptera. *Monitore Zoologico Italiano*. (*N.S.*), Supplement, **5**: 221–298.

Yalden, D.W., Largen, M.J. & Kock, D. 1976. Catalogue of the mammals of Ethiopia. 2. Insectivora and Rodentia. *Ibid*. Supplement, **8**: 1–118; 1977. 3. Primates. *Ibid*. Supplement, **9**. 1–52, 1980. 4. Carnivora. *Ibid*. Supplement, **13**: 169–272. 1984. 5. Artiodactyla. *Ibid*. Supplement, **19**: 67–221.

Meester, J. & Setzer, H.W. (Eds.) 1971. *The mammals of Africa: an identification manual*. Washington: Smithsonian Institution. (Includes keys to all species.)

Dorst, J. & Dandelot, P. 1970. *A field guide to the larger mammals of Africa*. London: Collins, 287 pp.

North and Central America, West Indies

Zyll de Jong, C.G. van 1983. *Handbook of Canadian mammals. I. Marsupials and insectivores*. Ottawa: National Museum of Natural Sciences, 210 pp.

Wilson, D.E. 1983. Checklist of mammals. Pp. 443–447 in Janzen, D.H. (Ed.) *Costa Rica natural history*. Chicago: University of Chicago Press, 816 pp.

Jones, J.K. *et al.* 1982. Revised checklist of North American mammals north of Mexico, 1982. *Occasional Papers, Museum Texas Tech University*, **80**: 22 pp.

Chapman, J.A. & Feldhammer, G.A. 1982. *Wild mammals of North America: biology, management and economics*. Baltimore etc: Johns Hopkins Press, 1184 pp.

Pulido, J.R. *et al.* 1982. *Catalogo de los mamiferos terrestres nativos de Mexico*. Mexico: Ed. Trillas, 126 pp.

Hall, E.R. 1981. *The mammals of North America*, 2nd edn. New York: Wiley, xv + 1181 + 90 pp, 2 vols. (Includes Central America south to Panama and the West Indies.)

Banfield, A.W.F. 1974. *The mammals of Canada*. Toronto.

South America

Husson, A.M. 1978. *The mammals of Surinam*. Leiden: Brill, xxxiv + 569 pp, 161 pls.

Peterson, N.E. & Pine, R.H. 1982. [Key for identification of the mammals of the Brazilian Amazon region with the exception of bats and primates]. *Acta Amazonica*, **12**: 465–482. (Portuguese, English summary.)

Orlog, C.C. & Lucero, M.M. 1981. *Guia de los mammiferos Argentinos*. Tucuman, 151 pp.

Mann Fischer, G. 1978. Los pequinos mamiferos de Chile (marsupiales, quiropteros, edentados y roedores). *Guyana* (Zool.), **40**: 342 pp.

Handley, C.O. 1976. Mammals of the Smithsonian Venezuelan Project. *Brigham Young University Science Bulletin*, Biol. Ser. **20** (5): 1–89. (All species recorded from Venezuela are mentioned.)

Ximenez, A., Langguth, A. & Praderi, R. 1972. Lista sistematica de los mamiferos del Uruguay. *Anales del Museo de Historia Natural de Montevideo*, **7** (5): 1–49.

Cabrera, A. 1957, 1961. Catalogo de los mamiferos de America del Sur. *Revista del Museo Argentino de Ciencias Naturales 'Bernardino Rivadavia'*, Ser. Cienc. zool. **4**, 732 pp. (A detailed nomenclatorial checklist.)

Taxonomic sources

The following sources have been used to update the basic geographical sources in compiling the list. They are quoted in the text of the list and are arranged here approximately in the sequence in which they appear in the list.

1. Monotremata, Marsupialia

1.1 Kirsch, J.A.W. & Calaby, J.H. 1977. The species of living marsupials: an annotated list. *In* Stonehouse, B. & Gilmore, D. (eds.) *The biology of marsupials.* London etc.: Macmillan: 9–26.

1.2 Archer, M. 1982. *Carnivorous marsupials.* 2 vols. Mosman, NSW, Australia: Royal Zoological Society of NSW, 804 pp.

1.3 Pine, R.H. & Handley, C.O. 1984. A review of the Amazonian short-tailed opossum *Monodelphis emiliae* (Thomas). *Mammalia,* **48**: 239–245.

1.4 Massoia, E. 1980. Un marsupial nuevo para la Argentina: *Monodelphis scalops . . . Physis,* **39**, **C**: 61–66.

1.5 Kitchener, D.J., Stoddart, J. & Henry, J. 1983. A taxonomic appraisal of the genus *Ningaui* Archer . . . including description of a new species. *Australian Journal of Zoology,* **31**: 361–379.

1.6 Kitchener, D.J., Stoddart, J. & Henry, J. 1984. A taxonomic revision of the *Sminthopsis murina* complex . . . in Australia . . . *Record of the Western Australian Museum,* **11**: 201–248.

1.7 McKenzie, N.L. & Archer, M. 1982. *Sminthopsis youngsoni* (Marsupialia: Dasyuridae) the lesser hairy-footed dunnart, a new species from arid Australia. *Australian Mammalogy,* **5** (4): 267–279.

1.8 Lyne, A.G. & Mort, P.A. 1981. A comparison of skull morphology in the marsupial bandicoot genus *Isoodon . . . Australian Mammalogy,* **4**: 107–133.

1.9 McKay, G.M. 1982. Nomenclature of the gliding possum genera *Petaurus* and *Petauroides . . . Australian Mammalogy,* **5**: 37–39.

1.10 Maynes, G.M. 1982. A new species of rock wallaby, *Petrogale persephone . . .* from Prosperine, central Queensland. *Australian Mammalogy,* **5**: 47–58.

1.11 Groves, C.P. 1982. The systematics of tree kangaroos (*Dendrolagus . . .*). *Australian Mammalogy,* **5**: 157–186.

1.12 Pine, R.H. 1981. Review of the mouse opossums *Marmosa parvidens* Tate and *Marmosa invicta* Goldman . . . *Mammalia,* **45**: 55–70.

1.13 Van Dyck, 1980. The cinnamon antechinus, *Antechinus leo . . .* a new species from the vine-forests of Cape York Peninsula. *Australian Mammalogy,* **3**: 5–17.

1.14 Ziegler, A.C. 1981. *Petaurus abidi,* new species of glider . . . from Papua New Guinea. *Australian Mammalogy,* **4**: 81–88.

1.15 Seebeck, J.H. & Johnston, P.G. 1980. *Potorous longipes,* new species . . . from eastern Victoria, Australia. *Australian Journal of Zoology,* **28**: 119–134.

1.16 Archer, M.A. 1982. Review of the dasyurid (Marsupialia) fossil record, . . . Pp. 397–443 *in* Archer, M. (Ed.) *Carnivorous marsupials.* Mosman, New South Wales: Royal Zoological Society of New South Wales.

2. Edentata

2.1 Wetzel, R.M. 1982. Systematics, distribution, ecology and conservation of South American edentates. Pp. 345–375 *In* Mares, M.A. & Genoways, H.H. *Mammalian biology in South America.* Special Publications. Pymatuning Laboratory of Ecology. University of Pittsburgh, Vol. 6.

2.2 Wetzel, R.M. 1980. Revision of the naked-tailed armadillos, genus *Cabassous* McMurtie. *Annals of the Carnegie Museum,* **49**: 323–357.

3. Insectivora

3.1 Poduschka, W. & Poduschka, C. 1982. Die taxonomische Zugehorigkeit von *Dasogale fontoynonti* G. Grandidier, 1928. *Sitzungsberichte der Osterreichischen Akademie der Wissenschaften. Mathematisch-Naturwissenschaftliche Klasse*, **191**: 253–264.

3.2 Van Valen, L. 1967. New Palaeocene insectivores and insectivore classification. *Bulletin of the American Museum of Natural History*, **135**: 219–284.

3.3 Heaney, C.R. & Morgan, G.S. 1982. A new species of gymnure, *Podogymnura*, (Mammalia, Erinaceidae) from Dinagat Island, Philippines. *Proceedings of the Biological Society of Washington*, **95**: 13–26.

3.4 Junge, J.A. & Hoffmann, R.S. 1981. An annotated key to the long-tailed shrews (genus *Sorex*) of the United States and Canada, with notes on Middle American *Sorex*. *Occasional Papers of the Museum of Natural History, University of Kansas, Lawrence*, **94**: 48 pp.

3.5 Hoffmann, R.S. (in press) A review of the genus *Soriculus* . . . .

3.6 Jenkins, P.D. 1982. A discussion of Malayan and Indonesian shrews of the genus *Crocidura* . . . *Zoologische Mededeelingen*, **56**: 267–279.

3.7 Hutterer, R. 1981. *Crocidura manengubae* n. sp. . . . eine neue Spitzmaus aus Kamerun. *Bonner Zoologische Beitrage*, **32**: 241 248.

3.8 Hutterer, R. 1983. Status of some African *Crocidura* described by Isidore Geoffroy Saint-Hilaire, Carl J. Sundevall and Theodor von Heuglin. *Annales, Musée Royal de l'Afrique Centrale, Sciences Zoologiques*, **237**: 207–217.

3.9 Hutterer, R. 1983. Taxonomy and distribution of *Crocidura fuscomurina* (Heuglin, 1863). *Mammalia*, **47**: 221 227.

3.10 Hutterer, R. & Kock, D. 1983. Spitzmäuse aus den Nuba-Bergen Kordofans, Sudan. *Senckenbergiana Biologica*, **63**: 17–26.

3.11 Hutterer, R. 1983. *Crocidura grandiceps*, eine neue Spitzmaus aus Westafrika. *Revue Suisse de Zoologie*, **90**: 699–707.

3.12 Hutterer, R. & Happold, D.C.D. 1983. The shrews of Nigeria . . . *Bonner Zoologische Monographien*, **18**: 79 pp.

3.13 Jenkins, P.D. 1984. Description of a new species of *Sylvisorex* . . . from Tanzania. *Bulletin of the British Museum (Natural History) (Zoology)*, **47**: 65–76.

3.14 Hoffmann, R.S. 1984. A review of the shrew-moles (genus *Uropsilus*) of China and Burma. *Journal of the Mammalogical Society of Japan*, **10**: 69–80.

3.15 Stogov, I.I. 1985. On two little studied species of white-toothed shrews . . . from the mountain regions in the south of the USSR. *Zoologicheskii Zhurnal*, **64**: 264–268.

3.16 Dippenaar, N.J. 1980. New species of *Crocidura* from Ethiopia and Northern Tanzania . . . *Annals of the Transvaal Museum*, **32**: 125–154.

4. Chiroptera

4.1 Miller, G.S. 1907. The families and genera of bats. *Bulletin of the United States National Museum*, **57**: i–xvii, 1–282, 49 pls.

4.2 Jones, J.K., Swanepoel, P. & Carter, D.C. 1977. Annotated checklist of the bats of Mexico and Central America. *Occasional Papers. The Museum, Texas Tech University*, no. 47, 35 pp.

4.3 Barbour, R.W. & Davis, W.H. 1969. *Bats of America*. Lexington: University Press of Kentucky.

4.4 Villa-R., B. 1966. *Los murcielagos de Mexico*. Mexico: Instituto de Biologia, Universidad Nacional Autónoma de Mexico.

4.5 Husson, A.M. 1962. The bats of Suriname. *Zoologische Verhandelingen, Leiden*, no. 58, 282 pp., 30 pls.

4.6 Goodwin, G.G. & Greenhall, A.M. 1961. A review of the bats of Trinidad and Tobago. *Bulletin of the American Museum of Natural History*, **122**: 187–302, 40 pls.

4.7 Baker, R.J. & Genoways, H.H. 1978. Zoogeography of Antillean bats. *In* Zoogeography in the Caribbean. *Special Publications. Academy of Natural Sciences of Philadelphia*, **13**: 53–97.

4.8 Brosset, A. 1959. The bats of Central and Western India. *Journal of the Bombay Natural History Society*, **59**: 1–57, 583–624, 707–746, 8 pls.

4.9 Andersen, K. 1912. *Catalogue of the Chiroptera in the collection of the British Museum. 1 Megachiroptera*. 2nd ed. London: British Museum (Natural History).

4.10 Koopman, K.F. 1980. Zoogeography of mammals from islands off the northern coast of New Guinea. *American Museum Novitates*, no. 2690: 17 pp.

4.11 Rookmaaker, L.C. & Bergmans, W. 1981. Taxonomy and geography of *Rousettus amplexicaudatus* (Geoffroy, 1810) ... *Beaufortia*, **31**: 1–29.

4.12 Hill, J.E. 1983. Bats ... from Indo-Australia. *Bulletin of the British Museum (Natural History)*, (Zool.), **45**: 103–208.

4.13 Bergmans, W & Hill, J.E. 1980. On a new species of *Rousettus* Gray, 1821, from Sumatra and Borneo ... *Bulletin of the British Museum (Natural History)*, (Zool.), **38**: 95–104.

4.14 Hill, J.E. & Francis, C.M. 1984. New bats ... and new records of bats from Borneo and Malaya. *Bulletin of the British Museum (Natural History)*, (Zool.), **47**: 305–329.

4.15 Bergmans, W. 1980. A new fruit bat of the genus *Myonycteris* Matschie, 1899, from eastern Kenya and Tanzania ... *Zoologische Mededeelingen, Leiden*, **55**: 171–181.

4.16 Musser, G.G., Koopman, K.F. & Caffia, D. 1982. The Sulawesian *Pteropus arquatus* and *P. argentatus* are *Acerodon celebensis*; the Philippine *P. leucotis* is an *Acerodon*. *Journal of Mammalogy*, **63**: 319–328.

4.17 Cheke, A.S. & Dahl, A.F. 1981. The status of bats on western Indian Ocean Islands, with special reference to *Pteropus*. *Mammalia*, **45**: 205–238.

4.18 Klingener, D. & Creighton, G.K. 1984. On small bats of the genus *Pteropus* from the Philippines. *Proceedings of the Biological Society of Washington*, **97**: 395–403.

4.19 Bergmans, W. & Sarbini, S. 1985. Fruit bats of the genus *Dobsonia* Palmer, 1898 from the islands of Biak, Owii, Numfoor and Yapen, Irian Jaya ... *Beaufortia*, **34**: 181–189.

4.20 Yenbutra, S. & Felten, H. 1983. A new species of the fruit bat genus *Megaerops* from SE-Asia. *Senckenbergiana Biologica*, **64**: 1–11.

4.21 Smith, J.D. & Hood, C.S. 1983. A new species of tube-nosed fruit bat (*Nyctimene*) from the Bismarck Archipelago, Papua New Guinea. *Occasional Papers, The Museum, Texas Tech University*, no. **81**: 1–14.

4.22 Heaney, L.R. & Peterson, R.L. 1984. A new species of tube-nosed fruit bat (*Nyctimene*) from Negros island, Philippines ... *Occasional Papers of the Museum of Zoology, University of Michigan*, no. **708**: 1–16.

4.23 Rozendaal, F.G. 1984. Notes on macroglossine bats from Sulawesi and the Moluccas, Indonesia ... *Zoologische Mededeelingen, Leiden*, **58**: 187–212.

4.24 Goodwin, R.E. 1978. The bats of Timor: systematics and ecology. *Bulletin of the American Museum of Natural History*, **163**: 73–122.

4.25 Ziegler, A.C. 1982. The Australo-Papuan genus *Syconycteris* ... *Occasional Papers of the Bernice Pauahi Bishop Museum*, **25** (5): 1–22.

4.26 Smith, J.D. & Hood, C.S. 1981. Preliminary notes on bats from the Bismarck Archipelago ... *Science in New Guinea*, **8**: 81–121.

4.27 Kitchener, D.J. 1980. *Taphozous hilli* sp. nov. ... a new sheath-tailed bat from Western Australia and Northern Territory. *Record of the Western Australian Museum*, **8**: 161–169.

4.28 McKean, J.L. & Friend, G.R. 1979. *Taphozous kapalgensis*, a new species of sheath-tailed bat from the Northern Territory, Australia. *Victoria Naturalist*, **96**: 239–241.

4.29 Kock, D. 1981. Zwei Fledermäuse neu für Kenya ... *Senckenbergiana Biologica*, (1980), **61**: 321–327.

4.30 Bohme, W. & Hutterer, R. 1978. Kommentierte Liste einer Säugetier-Aufsammlung aus dem Senegal. *Bonner Zoologische Beitrage,* **29**: 303–322.

4.31 Hill, J.E. & Yoshiyuki, M. 1980. A new species of *Rhinolophus* ... from Iriomote Island, Ryukyu Islands ... *Bulletin of the National Science Museum, Tokyo,* Ser. A. (Zool.), **6**: 179–189.

4.32 Smith, J.D. & Hill, J.E. 1981. A new species and subspecies of bat of the *Hipposideros bicolor* – group from Papua New Guinea ... *Contributions in Science,* **331**: 1–19.

4.33 Jenkins, P.D. & Hill, J.E. 1981. The status of *Hipposideros galeritus* Cantor, 1846 and *Hipposideros cervinus* (Gould, 1854) ... *Bulletin of the British Museum* (*Natural History*), (Zool.), **41**: 279–294.

4.34 Khajuria, H. 1970. A new leaf-nosed bat from Central India. *Mammalia,* **34**: 622–627.

4.35 Khajuria, H. 1982. External genitalia and bacula of some central Indian Microchiroptera. *Säugetierkundliche Mitteilungen,* **30**: 287–295.

4.36 Hill, J.E. & Venbutra, S. A new species of the *Hipposideros bicolor* group ... from Thailand. *Bulletin of the British Museum* (*Natural History*), (Zool.), **47**: 77–82.

4.37 Brosset, A. 1984. Chiroptères d'altitude du Mont Nimba (Guinée). Description d'une espèce nouvelle, *Hipposideros lamottei*. *Mammalia,* **48**: 544–555.

4.38 Hill, J.E. 1982. A review of the leaf-nosed bats *Rhinonycteris, Cloeotis* and *Triaenops*. *Bonner Zoologische Beitrage,* **33**: 165 186.

4.39 Silva Taboada, G. 1976. Historia y actualización taxónomica de algunes especies Antillanas de murciélagos de los géneros *Pteronotus, Brachyphylla, Lasiurus,* y *Antrozous* ... *Poeyana,* **153**: 1–24.

4.40 Hall, E.R. 1981. *The mammals of North America.* New York, etc: Wiley.

4.41 Davis, W.B. 1976. Notes on the bats *Saccopteryx canescens* Thomas and *Micronycteris hirsuta* (Peters). *Journal of Mammalogy,* **57**: 604–607.

4.42 Swanepoel, P. & Genoways, H.H. 1979. Morphometrics. *In* Baker, R.J., Jones, J.K. & Carter, D.C. (eds.) Biology of bats of the New World family Phyllostomatidae. Part III. *Special Publications, Museum, Texas Tech University,* **16**: 13–106.

4.43 Williams, S.L. & Genoways, H.H. 1980. Results of the Alcoa Foundation – Suriname Expeditions. II. Additional records of bats ... from Suriname. *Annals of the Carnegie Museum,* **49**: 213–236.

4.44 Ochoa, J. & Ibanez, C. 1984. Nuevo murcielago del genero *Lonchorhina*... *Memorias de la Sociedad de Ciencias Naturales 'La Salle'* (1982), **42** (118): 145–159

4.45 Gardner, A.L. 1976. The distributional status of some Peruvian mammals. *Occasional Papers of the Museum of Zoology, Louisiana State University,* **48**: 1–18.

4.46 Genoways, H.H. & Williams, S.L. 1980. Results of the Alcoa Foundation – Suriname Expeditions. I. A new species of bat of the genus *Tonatia*. *Annals of the Carnegie Museum,* **49**: 203–211.

4.47 Webster, W.D. & Jones, J.K. 1980. Taxonomic and nomenclatorial notes on bats of the genus *Glossophaga* in North America, with description of a new species. *Occasional Papers, The Museum, Texas Tech University,* **71**: 1–12.

4.48 Taddei, V.A., Vizotto, L.D. & Sazima, I. 1982. Una nova espéce de *Lonchophylla* do Brasil e clave para identifição das espécies do gênero ... *Ciencia et Cultura,* **35**: 625–629.

4.49 Hill, J.E. 1980. A note on *Lonchophylla* ... from Ecuador and Peru, with the description of a new species. *Bulletin of the British Museum* (*Natural History*), (Zool.), **38**: 233–236.

4.50 Handley, C. 1984. New species of mammals from northern South America: a long-tongued bat, genus *Anoura* Gray. *Proceedings of the Biological Society of Washington,* **97**: 513–521.

4.51 Davis, W.B. 1980. New *Sturnira* .... from Central and South America with key to currently recognised species. *Occasional Papers, The Museum, Texas Tech University,* no. 70: 1–5.

4.52 Anderson, S., Koopman, K.F. & Creighton, G.K. 1982. Bats of Bolivia: an annotated checklist. *American Museum Novitates*, no. 2750: 1–24.

4.53 Handley, C.D. 1976. Mammals of the Smithsonian Venezuelan Project. *Brigham Young University Science Bulletin*, Biol. Ser. 20, (5): 1–89, map.

4.54 Greenbaum, I.F., Baker, R.J. & Wilson, D.E. 1975. Evolutionary implications of the karyotypes of the stenodermine genera *Ardops, Ariteus, Phyllops*, and *Ectophylla*. *Bulletin of the Southern California Academy of Sciences*, **74**: 156–159.

4.55 Jones, J.K. & Carter, D.C. 1979. Systematic and distributional notes. *In* Baker, R.J., Jones, J.K. & Carter, D.C. (Eds.) Biology of bats of the New World family Phyllostomatidae. Part III. *Special Publications, Museum, Texas Tech University*, no. 16: 7–11.

4.56 Koopman, K.F. 1978. Zoogeography of Peruvian bats with special emphasis on the role of the Andes. *American Museum Novitates*, no. 2651: 1–33.

4.57 Davis, W.B. 1984. Review of the large fruit-eating bats of the *Artibeus 'lituratus'* complex . . . in Middle America. *Occasional Papers, The Museum, Texas Tech University*, no. 93: 1–16.

4.58 Swanepoel, P. & Genoways, H.H. 1978. Revision of the Antillean bats of the genus *Brachyphylla* . . . *Bulletin of Carnegie Museum of Natural History*, no. 12, 53 pp.

4.59 Koopman, K.F. 1982. In Honacki, J.H., Kinman, K.E. & Koeppl, J.W. *Mammal species of the world. A taxonomic and geographical reference*. Lawrence, Kansas: Allen Press Inc./Association of Systematics Collections.

4.60 Ruschi, A. 1951. Morcegos do Estado do Espirito Santo. Familia Vespertilionidae, . . . *Boletim do Museu de Biologia 'Prof. Mello-Leitao'*. (Zool.), no. 4: 1–11.

4.61 Pine, R.H. & Ruschi, A. 1976. Concerning certain bats described and recorded from Espirito Santo, Brazil. *Anales del Instituto de Biologia, Universidad de Mexico*, **47**, Ser. Zool. (2): 183–196.

4.62 Horáček, I. & Hanák, V. 1984. Comments on the systematics and phylogeny of *Myotis nattereri* (Kuhl, 1818). *Myotis*, **21–22**: 20–29.

4.63 Bogan, M.A. 1978. A new species of *Myotis* from the Islas Tres Marias, Nayarit, Mexico . . . *Journal of Mammalogy*, **59**: 519–530.

4.64 Hanák, V. & Horáček, I. 1984. Some comments on the taxonomy of *Myotis daubentoni* (Kuhl, 1819) *Myotis*, **21–22**: 7–19.

4.65 Dolan, P.G. & Carter, D.C. 1979. Distributional notes and records for Middle American Chiroptera. *Journal of Mammalogy*, **60**: 644–649.

4.66 Smithers, R.H.N. 1983. *The mammals of the Southern African subregion*. University of Pretoria, 736 pp.

4.67 Palmeirim, J.M. 1979. First record of *Myotis myotis* on the Azores Islands . . . *Archivos do Museu Bocage*. (2) *Notas e suplementos*, 7, no. 46: 1–2.

4.68 Genoways, H.H. & Williams, S.L. 1979. Notes on bats . . . from Bonaire and Curaçao, Dutch West Indies. *Annals of the Carnegie Museum*, **48**: 311–321.

4.69 Kobayashi, T., Maeda, K. & Harada, M. 1980. Studies on the small mammal fauna of Sabah, East Malaysia. I. Order Chiroptera and genus *Tupaia*. *Contributions from the Biological Laboratory Kyoto University*, **26**: 67–82.

4.70 Yoshiyuki, M. 1984. A new species of *Myotis* . . . from Hokkaido, Japan. *Bulletin of the National Science Museum, Tokyo*, Ser.A (Zool.), **10**(3): 153–158.

4.71 Menu, H. 1984. Révision du statut de *Pipistrellus subflavus* (F. Cuvier, 1832). Proposition d'un taxon generique nouveau: *Perimyotis* nov. gen. *Mammalia*, **48**: 409–416.

4.72 Koopman, K.F. 1973. Systematics of Indo-Australian *Pipistrellus*. *Periodicum Biologorum*, **75**: 113–116.

4.73 Harrison, D.L. 1979. A new species of pipistrelle bat (*Pipistrellus*: Vespertilionidae) from Oman. *Mammalia*, **43**: 573–576.

4.74 Robbins, C.B. 1980. Small mammals from Togo and Benin. I. Chiroptera. *Mammalia*, **44**: 83–88.

4.75 Hanák, V. & Gaisler, J. 1983. *Nyctalus leisleri* (Kuhl, 1818), une espèce nouvelle pour le continent africain. *Mammalia*, **47**: 585–587.

4.76 Jones, G.S. 1983. Ecological and distributional notes on mammals from Vietnam, including the first record of *Nyctalus. Mammalia*, **47**: 339–344.

4.77 Kock, D. 1981. Zur Chiropteren-Fauna von Burundi (Mammals). *Senckenbergiana Biologica*, (1980), **61**: 329–336.

4.78 Hill, J.E. & Evans, P.G.H. 1985. A record of *Eptesicus fuscus . . .* from Dominica, West Indies. *Mammalia*, **49**: 133–136.

4.79 DeBlase, A.F. 1980. The bats of Iran: systematics, distribution, ecology. *Fieldiana, Zoology*, (N.S.), no. 4 (Pub. 1307): i–xvii, 1–424.

4.80 Kock, D. 1981. *Philetor brachypterus* auf Neu-Britannien und dem Philippinen . . . *Senckenbergiana Biologica*, (1980), **61**: 313–319.

4.81 Koopman, K.F. 1983. A significant range extension of *Philetor . . .* with remarks on geographical variation. *Journal of Mammalogy*, **64**: 525–526.

4.82 Peterson, R.F. 1982. A new species of *Glauconycteris* from the east coast of Kenya . . . *Canadian Journal of Zoology*, **60**: 2521–2525.

4.83 Koopman, K.F. 1971. Taxonomic notes on *Chalinolobus* and *Glauconycteris*. *American Museum Novitates*, no. 2451: 1–10.

4.84 Hill, J.E. 1974. A review of *Scotoecus . . . Bulletin of the British Museum (Natural History)*, (Zool.), **27**: 167–188.

4.85 Kitchener, D.J. & Caputi, N. 1985. Systematic revision of Australian *Scoteanax* and *Scotorepens . . . Record of the Western Australian Museum*, **12**: 85–146.

4.86 Baker, R.J. 1984. A sympatric cryptic species of mammal: a new species of *Rhogeessa . . . Systematic Zoology*, **33**: 178–183.

4.87 Hill, J.E. 1980. The status of *Vespertilio borbonicus* E. Geoffroy, 1803 . . . *Zoologische Mededeelingen, Leiden*, **55**: 281–295.

4.88 Silva Taboada, G. 1979. *Los murcielagos de Cuba*. Editorial Academica: Havana, xiv + 423 pp.

4.89 Koopman, K.F. 1984. Taxonomic and distributional notes on tropical Australian bats. *American Museum Novitates*, no. 2778: 1–48.

4.90 Peterson, R.L. 1981. Systematic variation in the *tristis* group of the bent-winged bats of the genus *Miniopterus. Canadian Journal of Zoology*, **59**: 828–843.

4.91 Maeda, K. 1982. Studies on the classification of *Miniopterus* in Eurasia, Australia and Melanesia. *Honyurui Kagaku (Mammalian Science)*, Suppl. no. 1: 1–176.

4.92 Peterson, R.L. 1981. The systematic status of *Miniopterus australis* and related forms. *Bat Research News*, **22**: 48.

4.93 Melville, D.S. 1983. Notes on small mammals from Chiang Mai Province, including two species new to Thailand. *Natural History Bulletin of the Siam Society*, **31**: 157–162.

4.94 Yoshiyuki, M. 1983. A new species of *Murina* from Japan . . . *Bulletin of the National Science Museum, Tokyo*, Ser.A (Zool.), **9**: 141–148.

4.95 Engstrom, M.D. & Wilson, D.E. 1981. Systematics of *Antrozous dubiaquercus . . .* with comments on the status of *Bauerus* Van Gelder. *Annals of the Carnegie Museum*, **50**: 371–383.

4.96 Hill, J.E. & Koopman, K.F. 1981. The status of *Lamingtona lophorhina. Bulletin of the British Museum (Natural History)*, (Zool.), **41**: 275–278.

4.97 Hill, J.E. & Pratt, 1981. A record of *Nyctophilus timoriensis . . .* from New Guinea. *Mammalia*, **45**: 264–266.

4.98 Daniel, M.J. & Williams, G.R. 1984. A survey of the distribution, seasonal activity and roost sites of New Zealand bats. *New Zealand Journal of Ecology*, **7**: 9–25.

4.99 Hill, J.E. & Daniel, M.J. 1985. Systematics of the New Zealand short-tailed bat *Mystacina*. *Bulletin of the British Museum (Natural History)*, (Zool.) **48**: 279–300.

4.100 Freeman, P.W. 1981. A multivariate study of the family Molossidae . . . *Fieldiana, Zoology*, N.S. no. 7 (Pub. 1316): i–vii, 1–173.

4.101 Legendre, S. 1984. Étude odontologique des réprésentants actuels du groupe *Tadarida* . . . *Revue Suisse de Zoologie*, **91**: 399–442.

4.102 El-Rayah, M.A. 1981. A new bat of the genus *Tadarida* . . . from West Africa. *Life Sciences Occasional Papers, Royal Ontario Museum*, no. 36: 1–10.

4.103 Ibáñez, C. 1980. Descripćion de un nuevo género de quiroptero Neotropical de la familia Molossidae. *Doñana Acta Vertebrata*, **7**: 104–111.

4.104 Williams, S.L. & Genoways, H.H. 1980. Results of the Alcoa Foundation – Suriname Expeditions. IV. A new species of bat of the genus *Molossops* . . . *Annals of the Carnegie Museum*, **49**: 487–498.

4.105 Carter, D.C. & Dolan, P.G. 1978. Catalogue of type specimens of Neotropical bats in selected European Museums. *Special Publications, Museum, Texas Tech University*, no. 15: 1–136.

4.106 Williams, S.L. & Genoways, H.H. 1980. Results of the Alcoa Foundation – Suriname Expeditions. II. Additional records of bats from Suriname. *Annals of the Carnegie Museum*, **49**: 213–236.

4.107 Zyll de Jong, C.G. van 1984. Taxonomic relationships of Nearctic small-footed bats of the *Myotis leibii* group . . . *Canadian Journal of Zoology*, **62**: 2519–2526.

4.108 Robbins, C.B., De Vree, F. & Van Cakenberghe, V. 1983. A review of the systematics of the African bat genus *Scotophilus* . . . *Annales. Musée Royal de l'Afrique Centrale*, (Sci. Zool.), 237: 25.

4.109 Robbins, C. 1983. A new high forest species in the African bat genus *Scotophilus* . . . *Annales. Musée Royal de l'Afrique Centrale*, (Sci. Zool.), 237: 19–24.

5. Primates

5.1 Napier, P.H. 1976, 1981, 1985. *Catalogue of primates in the British Museum (Natural History)* . . . *Part 1: families Callitrichidae and Cebidae: Part 2: family Cercopithecidae, subfamily Cercopithecinae: Part 3: family Cercopithecidae, subfamily Colobinae.* London: British Museum (Natural History).

5.2 Napier, J.R. & Napier, P.H. 1985. *Natural history of the primates.* London: British Museum (Natural History), 200 pp.

5.3 Wolfheim, J.H. 1983. *Primates of the world: distribution, abundance and conservation.* Chur (Switzerland): Harwood, 831 pp.

5.4 Tattersall, I. 1982. *The primates of Madagascar.* New York: Columbia University Press, xiv + 382 pp.

5.5 Rosenberger, A.L. & Coimbra-Filho, A.F. 1984. Morphology, taxonomic status and affinities of the lion tamarins, *Leontopithecus* . . . *Folia Primatologica*, **42**: 149–179.

5.6 Hershkovitz, P. 1983. Two new species of night monkeys, genus *Aotus* (Cebidae, Platyrrhini): a preliminary report on *Aotus* taxonomy. *American Journal of Primatology*, **4**: 209–243.

5.7 Hershkovitz, P. 1984. Taxonomy of squirrel monkeys genus *Saimiri* . . . *American Journal of Primatology*, **7**: 155–210.

5.8 Thys van den Audenaerde, D.F.E. 1977. Description of a monkey-skin from East-Central Zaire as a probably new monkey-species . . . *Revue de Zoologie Africaine*, **91**: 1000–1010.

5.9 Brandon-Jones, D. 1984. Colobus and leaf monkeys. Pp. 398–408 in Macdonald, D. (Ed.) *The encyclopaedia of mammals*, vol. 1. London: Allen & Unwin.

6. Carnivora

6.1 Crawford-Cabral, J. 1982. The classification of the genets . . . *Boletim da Sociedade Portuguesa de Ciencias Naturais*, **20**: 97–114.

6.2 Goldman, C.A. 1984. Systematic revision of the African mongoose genus *Crossarchus* . . . *Canadian Journal of Zoology*, **62**: 1618–1630.

6.3 Ridgeway, S.H. & Harrison, R.J. 1981. *Handbook of marine mammals. Vol. 1. The walrus, sea lions, fur seals and sea otters*, 235 pp. *Vol. 2. Seals*, 359 pp. London etc: Academic Press.

6.4 King, J.E. 1983. *Seals of the world*, 2nd ed. London; Oxford: British Museum (Natural History), Oxford University Press, 240 pp.

7. Cetacea

7.1 Baker, A.N. 1983. *Whales and dolphins of New Zealand and Australia: an identification guide*. Wellington: Victoria University Press, 133 pp.

7.2 Watson, L. 1981. *Sea guide to whales of the world*. London: Hutchinson, 302 pp.

7.3 Berzin, A.A. & Vladimirov, V.L. 1983. A new species of killer whale (Cetacea, Delphinidae) from the Antarctic waters. *Zoologicheskii Zhurnal*, **62**: 287 295.

8. Ungulates, Pholidota

8.1 Wetzel, R.M. *et al.* 1975. *Catagonus*, an 'extinct' peccary, alive in Paraguay. *Science*, **189**: 379–381.

8.2 Li, Z.-x. 1981. On a new species of musk deer from China. *Zoological Research*, **2**: 157–160 (Chinese), 161 (English).

8.3 Groves, C.P. & Grubb, P. 1982. The species of muntjak (genus *Muntiacus*) in Borneo . . . *Zoologische Mededeelingen. Leiden*, **56**: 203–214.

8.4 Grubb, P. & Groves, C.P. 1983. Notes on the taxonomy of the deer . . . of the Philippines. *Zoologischer Anzeiger*, **210**: 119–144.

8.5 Yalden, D.W. 1978. A revision of the dik-diks of the subgenus *Madoqua* (*Madoqua*). *Monitore Zoologico Italiano*, N.S. Suppl. **11**: 245–264.

8.6 Patterson, B. 1979. Pholidota and Tubulidentata. *In* Maglio, V.J. & Cooke, H.B.S. (Eds.) *Evolution of African mammals*. Harvard.

8.7 Groves, C.P. & Lay, D.M. 1985. A new species of the genus *Gazella* . . . from the Arabian Peninsula. *Mammalia*, **49**: 27–36.

9. Rodentia

9.1 Cao Van Sung 1984. Inventaire des rongeurs du Vietnam. *Mammalia*, **48**: 391–395.

9.2 Watts, C.H.S. & Aslin, H.J. 1981. *The rodents of Australia*. London etc: Angus & Robertson, 321 pp.

9.3 De Graff, G. 1981. *The rodents of Southern Africa*. Durban etc: Butterworths, 267 pp.

9.4 Menzies, J.I. & Dennis, E. 1979. *Handbook of New Guinea rodents*. Wau, New Guinea: Wau Ecology Institute, 68 pp.

9.5 Patterson, B.D. 1984. Geographic variation and taxonomy of Colorado and Hopi chipmunks (genus *Eutamias*). *Journal of Mammalogy*, **65**: 442–456.

9.6 Levenson, H. & Hoffmann, R.S. 1984. Systematic relationships among taxa in the Townsend chipmunk group. *Southwestern Naturalist*, **29**: 157–168.

9.7 Saha, S.S. 1981. A new genus and a new species of flying squirrel (Mammalia: Rodentia: Sciuridae) from northwestern India. *Bulletin of the Zoological Survey of India*, **4**(3): 331–336.

9.8 Chakraborty, S. 1981 Studies on *Sciuropterus baberi* Blyth . . . *Proceedings of the Zoological Society*. Calcutta, **32**: 57–63.

9.9 Rogers, D.S. & Schmidly, D.J. 1982. Systematics of spiny pocket mice (genus *Heteromys*) . . . *Journal of Mammalogy*, **63**: 375–386.

9.10 Locks, M. 1981. Nova especie de *Oecomys* de Brasilia, DF, Brasil (Cricetidae, Rodentia). *Boletim do Museu Nacional do Rio de Janeiro*, NS, **300**: 7pp.

9.11 Barbour, D.B. & Humphrey, S.R. 1982. Status of the silver rice rat (*Oryzomys argentatus*). *Florida Scientist*, **45**: 112–116.

9.12 Hutterer, R. & Hirsch, U. 1979. Ein neuer *Nesoryzomys* von der Insel Fernandina, Galapagos . . . *Bonner Zoologische Beitrage*, **30**: 276–283.

9.13 Lee, M.R. & Schmidly, D.J. 1977. A new species of *Peromyscus* (Rodentia: Muridae) from Coahuila, Mexico. *Journal of Mammalogy*, **58**: 263–268.

9.14 Patterson, B.D. Gallardo, M.H. & Freas, K.E. 1984. Systematics of mice of the subgenus *Akodon* . . . in southern South America . . . *Fieldiana Zoology*, NS, **23**: 16 pp.

9.15 Pine, R.H. 1976. A new species of *Akodon* from Isla de los Estados, Argentina. *Mammalia*, **40**: 63–68.

9.16 De Santis, L.J. & Justo, E.R. 1980. *Akodon* (*Abrothrix*) *mansoensis* sp. nov., un nuevo 'raton lanoso' de la provincia de Rio Negro, Argentina . . . *Neotropica*, **26**: 121–127.

9.17 Massoia, E. 1979. Descripcion de un genero y especie nuevos, *Bibimys torresi* . . . *Physis Buenos Aires*, **38c** (95): 1–7.

9.18 Pearson, O.P. 1984. Taxonomy and natural history of some fossorial rodents of Patagonia, southern Argentina. *Journal of Zoology*, London, **202**: 225–237.

9.19 Dubost, G. & Petter, F. 1978. Une espece nouvelle de 'rat-pecheur' de Guyane francaise: *Daptomys oyapocka* sp. novo . . . *Mammalia*, **42**: 435–439.

9.20 Kovalskaya, Yu. & Sokolov, V.E. 1980. *Microtus evoronensis* sp. n. . . . from the lower Amur territory. *Zoologicheskii Zhurnal*, **59**: 1409–1416.

9.21 Malygin, V.M. 1983. [*Systematics of the common voles*]. Moscow: Izdat. Nauka, 208 pp.

9.22 Lay, D.M. 1983. Taxonomy of the genus *Gerbillus* . . . *Zeitschrift fur Säugetierkunde*, **48**: 329–354.

9.23 Dieterlen, F. & Rupp, H. 1978. *Megadendromus nikolausi*, gen. nov., sp. nov., ein neuer Nager aus Athiopien. *Zeitschrift fur Säugetierkunde*, **43**: 129–143.

9.24 Hubert, B. 1978. Revision of the genus *Saccostomus* . . . *Bulletin of Carnegie Museum of Natural History*, **6**: 48–52.

9.25 Wang, Y., Hu, J. & Chen, K. 1980. A new species of Murinae – *Vernaya foramena* sp. nov. *Acta Zoologica Sinica*, **26**: 393–397.

9.26 Van der Straeten, E. 1975. *Lemniscomys bellieri*, a new species of Muridae for the Ivory Coast. *Revue Zoologique Africaine*, **89**: 906–908.

9.27 Van der Straeten, E. 1980. A new species of *Lemniscomys* (Muridae) from Zambia. *Annals of the Cape Provincial Museums*, **13**: 55–62.

9.28 Mishra, A.C. & Dhanda, V. 1975. Review of the genus *Millardia* . . . *Journal of Mammalogy*, **5**: 76–80.

9.29 Abe, H. 1983. Variation and taxonomy of *Niviventer fulvescens* and notes on *Niviventer* group of rats in Thailand. *Journal of the Mammalogical Society of Japan*, **9**: 151–161.

9.30 Musser, G.G., Marshall, J.T. & Boeadi. 1979. Definition and contents of the Sundaic genus *Maxomys* (Rodentia, Muridae). *Journal of Mammalogy*, **60**: 592–606.

9.31 Musser, G.G. 1981. Results of the Archbold Expeditions. No. 105. Notes on the systematics of Indo-Malayan murid rodents, and descriptions of new genera and species from Ceylon, Sulawezi and the Philippines. *Bulletin of the American Museum of Natural History*, **68**: 225–334.

9.32 Musser, G.G. 1981. The giant rat of Flores and its relatives east of Borneo and Bali. *Bulletin of the American Museum of Natural History*, **169**: 71–175.

9.33 Taylor, J.M. Calaby, J.H. & Van Deusen, H.M. 1982. A revision of the genus *Rattus* (Rodentia, Muridae) in the New Guinea Region. *Bulletin of the American Museum of Natural History*, **173**: 177–336.

9.34 Musser, G.G. & Newcomb, C. 1983. Malaysian murids and the giant rat of Sumatra. *Bulletin of the American Museum of Natural History*, **174**: 327–598.

9.35 Musser, G.G. 1982. Results of the Archbold Expeditions. No. 107. A new genus of arboreal rat from Luzon Island in the Philippines. *American Museum Novitates*, **2730**: 23 pp.

9.36 Musser, G.G. 1981. A genus of arboreal rat from west Java, Indonesia. *Zoologische Verhandelingen. Leiden*, **189**: 35 pp, 4 pls.

9.37 Musser G.G. 1982. Results of the Archbold Expeditions. No. 108. The definition of *Apomys*, a native rat of the Philippine Islands. *American Museum Novitates*, **2746**: 43 pp.

9.38 Van der Straeten, E. & Verheyen, W.N. 1978. Taxonomic notes on the West-African *Myomys* with the description of *Myomys derooi*. *Zeitschrift fur Säugetierkunde*, **43**: 31–41.

9.39 Capanna, E., Civitelli, M.V. & Ceraso, A. 1982. Karyotypes of Somalian rodent populations. 3. *Mastomys huberti*. *Monitore Zoologico Italiano, Supplement*. **16**: 141–152.

9.40 Van der Straeten, E. & Verheyen, W.N. 1981. Étude biométrique du genre *Praomys* en Cote d'Ivoire. *Bonner Zoologische Beitrage*. **32**: 249–264.

9.41 Van der Straeten, E. & Dieterlen, F. 1983. Description de *Praomys ruppi* une nouvelle espece de Muridae d'Ethiopie. *Annales. Musee Royal de l'Afrique Centrale*, Sc. Zool. **237**: 121–127.

9.42 Van der Straeten, E. 1984. Étude biométrique des genres *Dephomys* et *Stochomys* avec quelques notes taxonomiques . . . *Revue Zoologique Africaine*, **98**: 771–798.

9.43 Watts, C.H. 1976. *Leggadina lakedownensis*, a new species of murid rodent from North Queensland. *Transactions of the Royal Society of South Australia*, **100**: 105–108.

9.44 Kitchener, D.J. 1980. A new species of *Pseudomys* . . . from Western Australia. *Record of the Western Australian Museum*, **8**: 405–414.

9.45 Fox, B.J. & Briscoe, D. 1980. *Pseudomys pilligaensis*: a new species of murid rodent from the Pilliga Scrub, northern New South Wales. *Australian Mammalogy*, **3**: 109–126.

9.46 Winter, J.W. 1984. The Thornton Peak melomys, *Melomys hadrourus*, new species (Rodentia: Muridae) a new rainforest species from northeastern Queensland, Australia. *Memoirs of the Queensland Museum*, **21**: 519–540.

9.47 Vermeiren, L.J.P. & Verheyen, W.N. 1980. Notes sur les *Leggada* de Lamto, Cote d'Ivoire, avec la description de *Leggada baoulei* sp. n. . . . *Revue Zoologique Africaine*, **94**: 570–590.

9.48 Marshall, J.T. & Sage, R.D. 1981. Taxonomy of the house mouse. *Symposia of the Zoological Society of London*, **47**: 15–25.

9.49 Marshall, J. 1977. A synopsis of the Asian species of *Mus*. *Bulletin of the American Museum of Natural History*, **158**: 173–220.

9.50 Petter, F. 1978. Une souris nouvelle du sud de l'Afrique: *Mus setzeri* sp. nov. *Mammalia*, **42**: 377–379.

9.51 Musser, G.G. 1982. Results of the Archbold Expeditions. No. 110. *Crunomys* and the small-bodied shrew rats native to the Philippine Islands and Sulawezi (Celebes). *Bulletin of the American Museum of Natural History*, **174** (1): 1–95.

9.52 Spitzenberger, F. 1983. Die Stachelmaus von Kleinasien, *Acomys cilicicus* n. sp. *Annalen des Naturhistorischen Museums. Wien*, **81**: 443–446.

9.53 Harrison, D.L. 1980. The mammals obtained in Dhofar by the 1977 Oman Flora and Fauna Society. *In* The Scientific results of the Oman Flora and Fauna Survey 1977 (Dhofar). *Journal of Oman Studies*, Special Report 2: 387–397.

9.54 Petter, F. & Roche, J. 1981. Remarques preliminaires sur la systematique des *Acomys* (Rongeurs, Muridae). *Peracomys*, sous-genre nouveau. *Mammalia*, **45** (3): 381–383.

9.55 Khajuria, H. 1981. A bandicoot rat, *Erythronesokia bunnii*, new genus, new species . . . from Iraq. *Bulletin of the Natural History Research Center. Baghdad*, **7**: 157–164.

9.56 Dennis, E. & Menzies, J.I. 1979. A chromosomal and morphometric study of Papuan tree rats *Pogonomys* and *Chiruromys* (Rodentia, Muridae). *Journal of Zoology, London,* **189**: 315–332.

9.57 Wu, D. & Deng, X. 1984. A new species of tree mice from Yunnan, China. *Acta theriologica Sinica,* **4**: 207–21. (Chinese, English summary).

9.58 Musser, G.G. & Boeadi. 1980. A new genus of murid rodent from the Komodo Islands in Nusatenggara, Indonesia. *Journal of Mammalogy,* **61**: 395–413.

9.59 Musser, G.G. & Gordon, L.K. 1981. A new species of *Crateromys* . . . from the Philippines. *Journal of Mammalogy,* **62**: 513–525.

9.60 Musser, G.G., Gordon, L.K. & Sommer, H. 1982. Species-limits in the Philippine murid, *Chrotomys. Journal of Mammalogy,* **63**: 514–521.

9.61 Musser, G.G. & Piik, E. 1982. A new species of *Hydromys* from western New Guinea (Irian Jaya). *Zoologische Mededeelingen. Leiden,* **5**: 153–167.

9.62 Musser, G.G. & Freeman, P.W. 1981. A new species of *Rhynchomys* . . . from the Philippines. *Journal of Mammalogy,* **62**: 154–159.

9.63 Sokolov, V.E. *et al.* 1981. Revision of birch mice of the Caucasus: sibling species *Sicista caucasica* Vinogradov, 1925, and *S. kluchorica* sp. n. . . . *Zoologicheskii Zhurnal,* **60**: 1386–1393.

9.64 Sokolov, V.E. 1981. A new species of five-toed jerboa, *Allactaga nataliae* sp. n. . . . from Mongolia. *Zoologicheskii Zhurnal,* **60**: 793–795.

9.65 Vorontsov, N.N. & Shenbrot, G.I. 1984. A systematic review of the genus *Salpingotus* (Rodentia, Dipodidae), with a description of *Salpingotus pallidus* sp. n. from Kazakhstan. *Zoologicheskii Zhurnal,* **63** (5): 731–744.

9.66 Van Weers, D.J. 1983. Specific distinction in Old World porcupines. *Zoologische Garten,* **53**: 226–232.

9.67 Ximinez, A. 1981. Notas sobre el genero *Cavia* Pallas con la descripcion de *Cavia magna* sp. n. . . . *Revista Nordestina de Biologia,* 3 (special no.), 1980: 145–179.

9.68 Varona, L.S. 1979. Subgenero y especie nuevos de *Capromys* . . . *Poeyana,* **194**: 33 pp.

9.69 Kratochvil, J. *et al.* 1978. Capromyinae (Rodentia) of Cuba. 1. *Acta Scientiarum Naturalium Academiae Scientiarum Bohemoslovacae,* **12** (11): 60 pp.

9.70 Contreras, J. & Berry, L.M. 1982. *Ctenomys bonettoi,* una nueva especie de tuco-tuco de . . . Argentina . . . *Historia Natural, Mendoza,* **2** (14): 123–124.

9.71 Contreras, J. & Berry L.M. 1982. *Ctenomys argentinus,* una nueva especie de tuco-tuco de . . . Argentina . . . *Historia Natural, Mendoza,* **2** (20): 165–173.

9.72 Travi, V.H. 1981. Nota prévia sobre nova espéce do gênero *Ctenomys* . . . *Iheringia* Ser. Zool. **60**: 123–124.

9.73 Contreras, J.R. *et al.* 1977. *Ctenomys validus,* una nueva especie de 'tunduque' de la Provincia de Mendoza . . . *Physis, Buenos Aires,* C, **36** (92): 159–162.

9.74 Gardner, A.L. & Emmons, H. 1984. Species groups in *Proechimys* (Rodentia, Echimyidae) as indicated by karyology and bullar morphology. *Journal of Mammalogy,* **65** (1): 10–25.

9.75 Petter, F. 1978. Epidémiologie de la leishmaniose en Guyane francaise, en relation avec l'existance d'une espece nouvelle de rongeurs échimyidés, *Proechimys cuvieri* sp. n. *Compte Rendu Hebdomadaire des Seances de l'Academie des Sciences. Paris,* **287** (D): 261–264.

9.76 Avila-Pires, F.D. de & Wutke, M.R.C. 1981. Taxonomia e evolucao de *Clyomys* . . . *Revista Brasileira de Biologia,* **41**: 529–534.

9.77 Williams, D.F. & Mares, M.A. 1978. A new genus and species of phyllotine rodent from northwestern Argentina. *Annals of the Carnegie Museum,* **47**: 193–221.

9.78 Orlov, V.N. & Kovalskaya, Y.M. 1978. *Microtus mujanensis* sp. n. from the Vitim River Basin. *Zoologicheskii Zhurnal,* **57**: 1224–1232.

9.79 Massoia, E. 1980. Nuevos datos sobre *Akodon, Deltamys* y *Cabreramys* . . . *Historia Natural Mendoza,* **1** (25): 179.

9.80 Womochel, D.R. 1978. A new species of *Allactaga* . . . from Iran. *Fieldiana, Zoology,* **72**: 65–73.

9.81 Myers, P. & Carleton, M.D. 1981. The species of *Oryzomys* (*Oligoryzomys*) in Paraguay . . . *Miscellaneous Publications. Museum of Zoology, University of Michigan,* **61**: 41 pp.

9.82 Wang, Y. 1985. A new genus and species of Gliridae – *Chaetocauda sichuanensis* gen. et sp. nov. *Acta Theriologica Sinica,* **5**: 67–73 (Chinese); 74–75 (English).

9.83 Bates, P.J.J. 1985. Studies of gerbils of genus *Tatera* .... *Mammalia,* **49**: 37–52.

9.84 Musser, G.G. & Newcomb, C. 1985. Definitions of Indochinese *Rattus losea* and a new species from Vietnam. *American Museum Novitates* **2814,** 32 pp.

9.85 Musser, G.G. & Williams, M.M. 1985. Systematic studies of oryzomyine rodents . . . *American Museum Novitates* **2810,** 22 pp.

10. Lagomorpha, Macroscelidea

10.1 Corbet, G.B. 1983. A review of classification in the family Leporidae. *Acta Zoologica Fennica,* **174**: 11–15.

10.2 Green, J.S. & Flinders, J.T. 1980. *Brachylagus idahoensis. Mammalian Species,* **125**: 4 pp.

10.3 Corbet, G.B. & Hanks, J. 1968. A revision of the elephant-shrews, family Macroscelididae. *Bulletin of the British Museum* (*Natural History*), (Zool.), **16**: 47–111.

10.4 Wang, Y., Luo, Z. & Feng, Z. 1985. Taxonomic revision of Yunnan hare, *Lepus comus* G. Allen with description of two new subspecies. *Zoological Research,* **6**: 101–107 (Chinese), 108–109 (English).

Index